Good Laboratory Practice Regulations

DRUGS AND THE PHARMACEUTICAL SCIENCES

DRUGS AND THE PHARMACEUTICAL SCIENCES

A Series of Textbooks and Monographs

1. Pharmacokinetics, *Milo Gibaldi and Donald Perrier*
2. Good Manufacturing Practices for Pharmaceuticals: A Plan for Total Quality Control, *Sidney H. Willig, Murray M. Tuckerman, and William S. Hitchings IV*
3. Microencapsulation, *edited by J. R. Nixon*
4. Drug Metabolism: Chemical and Biochemical Aspects, *Bernard Testa and Peter Jenner*
5. New Drugs: Discovery and Development, *edited by Alan A. Rubin*
6. Sustained and Controlled Release Drug Delivery Systems, *edited by Joseph R. Robinson*
7. Modern Pharmaceutics, *edited by Gilbert S. Banker and Christopher T. Rhodes*
8. Prescription Drugs in Short Supply: Case Histories, *Michael A. Schwartz*
9. Activated Charcoal: Antidotal and Other Medical Uses, *David O. Cooney*
10. Concepts in Drug Metabolism (in two parts), *edited by Peter Jenner and Bernard Testa*
11. Pharmaceutical Analysis: Modern Methods (in two parts), *edited by James W. Munson*
12. Techniques of Solubilization of Drugs, *edited by Samuel H. Yalkowsky*
13. Orphan Drugs, *edited by Fred E. Karch*
14. Novel Drug Delivery Systems: Fundamentals, Developmental Concepts, Biomedical Assessments, *Yie W. Chien*
15. Pharmacokinetics: Second Edition, Revised and Expanded, *Milo Gibaldi and Donald Perrier*
16. Good Manufacturing Practices for Pharmaceuticals: A Plan for Total Quality Control, Second Edition, Revised and Expanded, *Sidney H. Willig, Murray M. Tuckerman, and William S. Hitchings IV*
17. Formulation of Veterinary Dosage Forms, *edited by Jack Blodinger*
18. Dermatological Formulations: Percutaneous Absorption, *Brian W. Barry*
19. The Clinical Research Process in the Pharmaceutical Industry, *edited by Gary M. Matoren*
20. Microencapsulation and Related Drug Processes, *Patrick B. Deasy*
21. Drugs and Nutrients: The Interactive Effects, *edited by Daphne A. Roe and T. Colin Campbell*
22. Biotechnology of Industrial Antibiotics, *Erick J. Vandamme*

23. Pharmaceutical Process Validation, *edited by Bernard T. Loftus and Robert A. Nash*
24. Anticancer and Interferon Agents: Synthesis and Properties, *edited by Raphael M. Ottenbrite and George B. Butler*
25. Pharmaceutical Statistics: Practical and Clinical Applications, *Sanford Bolton*
26. Drug Dynamics for Analytical, Clinical, and Biological Chemists, *Benjamin J. Gudzinowicz, Burrows T. Younkin, Jr., and Michael J. Gudzinowicz*
27. Modern Analysis of Antibiotics, *edited by Adjoran Aszalos*
28. Solubility and Related Properties, *Kenneth C. James*
29. Controlled Drug Delivery: Fundamentals and Applications, Second Edition, Revised and Expanded, *edited by Joseph R. Robinson and Vincent H. Lee*
30. New Drug Approval Process: Clinical and Regulatory Management, *edited by Richard A. Guarino*
31. Transdermal Controlled Systemic Medications, *edited by Yie W. Chien*
32. Drug Delivery Devices: Fundamentals and Applications, *edited by Praveen Tyle*
33. Pharmacokinetics: Regulatory • Industrial • Academic Perspectives, *edited by Peter G. Welling and Francis L. S. Tse*
34. Clinical Drug Trials and Tribulations, *edited by Allen E. Cato*
35. Transdermal Drug Delivery: Developmental Issues and Research Initiatives, *edited by Jonathan Hadgraft and Richard H. Guy*
36. Aqueous Polymeric Coatings for Pharmaceutical Dosage Forms, *edited by James W. McGinity*
37. Pharmaceutical Pelletization Technology, *edited by Isaac Ghebre-Sellassie*
38. Good Laboratory Practice Regulations, *edited by Allen F. Hirsch*
39. Nasal Systemic Drug Delivery, *Yie W. Chien, Kenneth S. E. Su, and Shyi-Feu Chang*
40. Modern Pharmaceutics: Second Edition, Revised and Expanded, *edited by Gilbert S. Banker and Christopher T. Rhodes*
41. Specialized Drug Delivery Systems: Manufacturing and Production Technology, *edited by Praveen Tyle*
42. Topical Drug Delivery Formulations, *edited by David W. Osborne and Anton H. Amann*
43. Drug Stability: Principles and Practices, *Jens T. Carstensen*
44. Pharmaceutical Statistics: Practical and Clinical Applications, Second Edition, Revised and Expanded, *Sanford Bolton*
45. Biodegradable Polymers as Drug Delivery Systems, *edited by Mark Chasin and Robert Langer*
46. Preclinical Drug Disposition: A Laboratory Handbook, *Francis L. S. Tse and James J. Jaffe*
47. HPLC in the Pharmaceutical Industry, *edited by Godwin W. Fong and Stanley K. Lam*
48. Pharmaceutical Bioequivalence, *edited by Peter G. Welling, Francis L. S. Tse, and Shrikant V. Dinghe*

49. Pharmaceutical Dissolution Testing, *Umesh V. Banakar*
50. Novel Drug Delivery Systems: Second Edition, Revised and Expanded, *Yie W. Chien*
51. Managing the Clinical Drug Development Process, *David M. Cocchetto and Ronald V. Nardi*
52. Good Manufacturing Practices for Pharmaceuticals: A Plan for Total Quality Control, Third Edition, *edited by Sidney H. Willig and James R. Stoker*
53. Prodrugs: Topical and Ocular Drug Delivery, *edited by Kenneth B. Sloan*
54. Pharmaceutical Inhalation Aerosol Technology, *edited by Anthony J. Hickey*
55. Radiopharmaceuticals: Chemistry and Pharmacology, *edited by Adrian D. Nunn*
56. New Drug Approval Process: Second Edition, Revised and Expanded, *edited by Richard A. Guarino*
57. Pharmaceutical Process Validation: Second Edition, Revised and Expanded, *edited by Ira R. Berry and Robert A. Nash*
58. Ophthalmic Drug Delivery Systems, *edited by Ashim K. Mitra*
59. Pharmaceutical Skin Penetration Enhancement, *edited by Kenneth A. Walters and Jonathan Hadgraft*
60. Colonic Drug Absorption and Metabolism, *edited by Peter R. Bieck*
61. Pharmaceutical Particulate Carriers: Therapeutic Applications, *edited by Alain Rolland*
62. Drug Permeation Enhancement: Theory and Applications, *edited by Dean S. Hsieh*
63. Glycopeptide Antibiotics, *edited by Ramakrishnan Nagarajan*
64. Achieving Sterility in Medical and Pharmaceutical Products, *Nigel A. Halls*
65. Multiparticulate Oral Drug Delivery, *edited by Isaac Ghebre-Sellassie*
66. Colloidal Drug Delivery Systems, *edited by Jörg Kreuter*
67. Pharmacokinetics: Regulatory • Industrial • Academic Perspectives, Second Edition, *edited by Peter G. Welling and Francis L. S. Tse*
68. Drug Stability: Principles and Practices, Second Edition, Revised and Expanded, *Jens T. Carstensen*
69. Good Laboratory Practice Regulations: Second Edition, Revised and Expanded, *edited by Sandy Weinberg*
70. Physical Characterization of Pharmaceutical Solids, *edited by Harry G. Brittain*
71. Pharmaceutical Powder Compaction Technology, *edited by Göran Alderborn and Christer Nyström*
72. Modern Pharmaceutics: Third Edition, Revised and Expanded, *edited by Gilbert S. Banker and Christopher T. Rhodes*
73. Microencapsulation: Methods and Industrial Applications, *edited by Simon Benita*
74. Oral Mucosal Drug Delivery, *edited by Michael J. Rathbone*
75. Clinical Research in Pharmaceutical Development, *edited by Barry Bleidt and Michael Montagne*

76. The Drug Development Process: Increasing Efficiency and Cost Effectiveness, *edited by Peter G. Welling, Louis Lasagna, and Umesh V. Banakar*
77. Microparticulate Systems for the Delivery of Proteins and Vaccines, *edited by Smadar Cohen and Howard Bernstein*
78. Good Manufacturing Practices for Pharmaceuticals: A Plan for Total Quality Control, Fourth Edition, Revised and Expanded, *Sidney H. Willig and James R. Stoker*
79. Aqueous Polymeric Coatings for Pharmaceutical Dosage Forms: Second Edition, Revised and Expanded, *edited by James W. McGinity*
80. Pharmaceutical Statistics: Practical and Clinical Applications, Third Edition, *Sanford Bolton*
81. Handbook of Pharmaceutical Granulation Technology, *edited by Dilip M. Parikh*
82. Biotechnology of Antibiotics: Second Edition, Revised and Expanded, *edited by William R. Strohl*
83. Mechanisms of Transdermal Drug Delivery, *edited by Russell O. Potts and Richard H. Guy*
84. Pharmaceutical Enzymes, *edited by Albert Lauwers and Simon Scharpé*
85. Development of Biopharmaceutical Parenteral Dosage Forms, *edited by John A. Bontempo*
86. Pharmaceutical Project Management, *edited by Tony Kennedy*
87. Drug Products for Clinical Trials: An International Guide to Formulation • Production • Quality Control, *edited by Donald C. Monkhouse and Christopher T. Rhodes*
88. Development and Formulation of Veterinary Dosage Forms: Second Edition, Revised and Expanded, *edited by Gregory E. Hardee and J. Desmond Baggot*
89. Receptor-Based Drug Design, *edited by Paul Leff*
90. Automation and Validation of Information in Pharmaceutical Processing, *edited by Joseph F. deSpautz*
91. Dermal Absorption and Toxicity Assessment, *edited by Michael S. Roberts and Kenneth A. Walters*
92. Pharmaceutical Experimental Design, *Gareth A. Lewis, Didier Mathieu, and Roger Phan-Tan-Luu*
93. Preparing for FDA Pre-Approval Inspections, *edited by Martin D. Hynes III*
94. Pharmaceutical Excipients: Characterization by IR, Raman, and NMR Spectroscopy, *David E. Bugay and W. Paul Findlay*
95. Polymorphism in Pharmaceutical Solids, *edited by Harry G. Brittain*
96. Freeze-Drying/Lyophilization of Pharmaceutical and Biological Products, *edited by Louis Rey and Joan C. May*
97. Percutaneous Absorption: Drugs–Cosmetics–Mechanisms–Methodology, Third Edition, Revised and Expanded, *edited by Robert L. Bronaugh and Howard I. Maibach*

98. Bioadhesive Drug Delivery Systems: Fundamentals, Novel Approaches, and Development, *edited by Edith Mathiowitz, Donald E. Chickering III, and Claus-Michael Lehr*
99. Protein Formulation and Delivery, *edited by Eugene J. McNally*
100. New Drug Approval Process: Third Edition, The Global Challenge, *edited by Richard A. Guarino*
101. Peptide and Protein Drug Analysis, *edited by Ronald E. Reid*
102. Transport Processes in Pharmaceutical Systems, *edited by Gordon L. Amidon, Ping I. Lee, and Elizabeth M. Topp*
103. Excipient Toxicity and Safety, *edited by Myra L. Weiner and Lois A. Kotkoskie*
104. The Clinical Audit in Pharmaceutical Development, *edited by Michael R. Hamrell*
105. Pharmaceutical Emulsions and Suspensions, *edited by Francoise Nielloud and Gilberte Marti-Mestres*
106. Oral Drug Absorption: Prediction and Assessment, *edited by Jennifer B. Dressman and Hans Lennernäs*
107. Drug Stability: Principles and Practices, Third Edition, Revised and Expanded, *edited by Jens T. Carstensen and C. T. Rhodes*
108. Containment in the Pharmaceutical Industry, *edited by James P. Wood*
109. Good Manufacturing Practices for Pharmaceuticals: A Plan for Total Quality Control from Manufacturer to Consumer, Fifth Edition, Revised and Expanded, *Sidney H. Willig*
110. Advanced Pharmaceutical Solids, *Jens T. Carstensen*
111. Endotoxins: Pyrogens, LAL Testing, and Depyrogenation, Second Edition, Revised and Expanded, *Kevin L. Williams*
112. Pharmaceutical Process Engineering, *Anthony J. Hickey and David Ganderton*
113. Pharmacogenomics, *edited by Werner Kalow, Urs A. Meyer, and Rachel F. Tyndale*
114. Handbook of Drug Screening, *edited by Ramakrishna Seethala and Prabhavathi B. Fernandes*
115. Drug Targeting Technology: Physical • Chemical • Biological Methods, *edited by Hans Schreier*
116. Drug–Drug Interactions, *edited by A. David Rodrigues*
117. Handbook of Pharmaceutical Analysis, *edited by Lena Ohannesian and Anthony J. Streeter*
118. Pharmaceutical Process Scale-Up, *edited by Michael Levin*
119. Dermatological and Transdermal Formulations, *edited by Kenneth A. Walters*
120. Clinical Drug Trials and Tribulations: Second Edition, Revised and Expanded, *edited by Allen Cato, Lynda Sutton, and Allen Cato III*
121. Modern Pharmaceutics: Fourth Edition, Revised and Expanded, *edited by Gilbert S. Banker and Christopher T. Rhodes*
122. Surfactants and Polymers in Drug Delivery, *Martin Malmsten*
123. Transdermal Drug Delivery: Second Edition, Revised and Expanded, *edited by Richard H. Guy and Jonathan Hadgraft*

124. Good Laboratory Practice Regulations: Second Edition, Revised and Expanded, *edited by Sandy Weinberg*
125. Parenteral Quality Control: Sterility, Pyrogen, Particulate, and Package Integrity Testing: Third Edition, Revised and Expanded, *Michael J. Akers, Daniel S. Larrimore, and Dana Morton Guazzo*
126. Modified-Release Drug Delivery Technology, *edited by Michael J. Rathbone, Jonathan Hadgraft, and Michael S. Roberts*
127. Simulation for Designing Clinical Trials: A Pharmacokinetic-Pharmacodynamic Modeling Perspective, *edited by Hui C. Kimko and Stephen B. Duffull*

ADDITIONAL VOLUMES IN PREPARATION

Pharmaceutical Process Validation: An International Third Edition, Revised and Expanded, *edited by Robert A. Nash and Alfred H. Wachter*

Affinity Capillary Electrophoresis in Pharmaceutics and Biopharmaceutics, *edited by Reinhard H. H. Neubert and Hans-Hermann Rüttinger*

Ophthalmic Drug Delivery Systems: Second Edition, Revised and Expanded, *edited by Ashim K. Mitra*

Pharmaceutical Gene Delivery Systems, *edited by Alain Rolland and Sean M. Sullivan*

Biomarkers in Clinical Drug Development, *edited by John Bloom*

Pharmaceutical Inhalation Aerosol Technology: Second Edition, Revised and Expanded, *edited by Anthony J. Hickey*

Good Laboratory Practice Regulations

Third Edition, Revised and Expanded

edited by
Sandy Weinberg

Muhlenberg College
Allentown, Pennsylvania, U.S.A.

MARCEL DEKKER
MARCEL DEKKER, INC. NEW YORK • BASEL

Library of Congress Cataloging-in-Publication Data
A catalog record for this book is available from the Library of Congress.

ISBN: 0-8247-0891-1

This book is printed on acid-free paper.

Headquarters
Marcel Dekker, Inc.
270 Madison Avenue, New York, NY 10016
tel: 212-696-9000; fax: 212-685-4540

Eastern Hemisphere Distribution
Marcel Dekker AG
Hutgasse 4, Postfach 812, CH-4001 Basel, Switzerland
tel: 41-61-260-6300; fax: 41-61-260-6333

World Wide Web
http://www.dekker.com

The publisher offers discounts on this book when ordered in bulk quantities. For more information, write to Special Sales/Professional Marketing at the headquarters address above.

Current printing (last digit):
10 9 8 7 6 5 4 3 2 1

PRINTED IN THE UNITED STATES OF AMERICA

Preface

The Good Laboratory Practices (GLPs) are extremely effective guidelines intended to provide management with a tool for controlling regulated laboratories and to provide regulators with a measurement guide for the evaluation of those controlling efforts. Over the past decade the GLPs have survived the test of field scrutiny and have evolved into a meaningful set of standards for a variety of laboratories worldwide.

The effects of evolution on regulation are generally positive. Vague requirements are clarified with example, dispute resolution, and dissemination of interpretation. Differences of opinion are debated, discussed, and synthesized. Guidelines are issued, investigators are trained, and examples are provided.

Working in conjunction with this delineation of regulation are the variances created by changes in practice, technology, and focus. The conjunction of these forces may create its own conflict, but experience and time have a soothing influence even on disagreement; positions and opinions may be at variance, but they are at least clear and specific.

The GLP regulations have reached this next stage of evolved understanding. The requirements are clear, the guidelines and interpretations are available, and the conflicts are resolved. Even the revolutionary influences of computerization in the laboratory have been measured and considered, providing the good automated laboratory practices corollary guidelines. The GLPs have come of age, and provide a clear and consistent framework for the assurance of quality and control in the laboratory.

This maturity has led to global GLPs that are, if not identical, at least compatible. In the United States, the same general guidelines apply to laboratories regulated by the Food and Drug Administration (FDA), Environmental Protection Agency (EPA) labs, and general analytical laboratories. Those same standards are consistent with required procedures in the European Community (EC); Switzerland, Japan, Israel, Brazil, and Russia. The Worldwide concern for quality and control had evolved the same general principles of operation and organization.

This third edition of *Good Laboratory Practice Regulations* reflects that evolutionary maturity of requirements. With a relative uniformity of interpretation more specific and applied information can be provided, furnishing the reader with both the theoretical overview necessary to anticipate new and emerging interpretations and the detail of practical information that can serve as a guide to the current standards and requirements of good laboratory management.

Chapter 1 provides a historical review of the evolutionary process, tracing the problems and concerns of regulators and the response of laboratory managers to those issues. The ultimate result, is the series of guidelines published as the GLPs, provided with a context and rationale.

Chapter 2 provides a definitive review of the current version of FDA GLPs, including an analysis and interpretation of the enforcement of the GLPs. This summary is an excellent overview for readers not previously familiar with the requirements.

Chapter 3 examines the new FDA regulation, Part 11, and its impact on a GLP laboratory setting. This chapter argues that Part 11 is a de facto extension of the GLPs and represents a historical update of the GLP regulation.

Chapters 4 and 5 focus on two important applications of the GLPs. In Chapter 4 the GLPs are applied to an automated laboratory as the (EPA) good automated laboratory practices. Chapter 5 examines the impact and interpretation of the GLPs in a non-GLP analytical laboratory. Stephanie Olexa makes an impressive case that the GLPs are relevant guidelines even where they are not regulatory requirements.

Chapter 6 provides specific standards and general guidelines for the validation of Laboratory Information Management Systems (LIMS) and other computerized laboratory systems. The chapter includes specific interpretations of the validation requirements in the post-Part 11 environment.

A review of standards without an examination of the interpretation and enforcement of those standards would be of little value. In Chapter 7 therefore, the FDA's GLP inspection program is dissected and analyzed, revealing the philosophy and approach of the regulators to GLP field interpretation.

In each previous edition the eighth and final chapter has been dedicated to the art of prediction, providing an extrapolation of GLP trends and applications into the near-term future. These predictions have provided gratifyingly accurate recommendations for anticipating new regulatory and practical changes. In this third edition, Chapter 8 looks at the next step in laboratory automation: robotic control of samples and analysis, the field laboratory, new laboratory applications in DNA and other genetic testing, and emerging new government perspectives on regulation and enforcement.

These eight chapters provide a detailed review of the GLP requirements, an examination of the application of those requirements, an interpretation of the effects of GLPs on an automated laboratory and the effects of that automation on the GLP guidelines, and a look at future trends in laboratories and their regulation.

Sandy Weinberg

Contents

Contributors

George W. James U.S. Food and Drug Administration, Rockville, Maryland, U.S.A.

Stephanie A. Olexa Benchmark Analytics, Center Valley, Pennsylvania, U.S.A.

Wendell A. Peterson Parke-Davis Pharmaceutical Research Division of Warner-Lambert Company, Ann Arbor, Michigan, U.S.A.

Paul L. Robinson Muhlenberg College, Allentown, Pennsylvania, U.S.A.

Gary C. Stein Weinberg, Spelton & Sax, Inc., Boothwyn, Pennsylvania, U.S.A.

Jean M. Taylor* U.S. Food and Drug Administration, Rockville, Maryland, U.S.A.

---

*Retired

Sandy Weinberg Muhlenberg College, Allentown, Pennsylvania, U.S.A.

Gerald J. Whartenby Muhlenberg College, Allentown, Pennsylvania, U.S.A.

Good Laboratory Practice Regulations

1

Historical Perspective

JEAN M. TAYLOR*
U.S. Food and Drug Administration, Rockville, Maryland, U.S.A.

GARY C. STEIN
Weinberg, Spelton & Sax, Inc., Boothwyn, Pennsylvania, U.S.A.

THE PROBLEM IN THE 1970s

FDA's Perspective

The Federal Food, Drug and Cosmetic Act (FFDCA) places the responsibility for establishing the safety and efficacy of human and veterinary drugs and devices and the safety of food and color additives on the sponsor of the regulated product. The Public Health Service Act requires that a sponsor

*Retired.

establish the safety and efficacy of biological products. These laws place on the Food and Drug Administration (FDA) the responsibility for reviewing the sponsor's test results and determining whether or not the results establish the safety and efficacy of the product. If the agency accepts that safety and efficacy are established adequately, the sponsor is permitted to market the product.

The types of scientific tests needed to establish safety are dependent on the nature of the regulated product and its proposed use. A product such as a food or color additive will require tests to elucidate the potential of the product to induce adverse acute, subchronic, and chronic effects. The safety tests are generally performed in animals and other biological systems. Both the types of tests and the methodology of particular tests have changed over the years with scientific advances in the field of toxicology.

The FDA regulations or guidelines prescribe the types of safety tests for a particular product. Sponsors may conduct the studies in their own laboratories or have them performed by a contract laboratory, a university, or some other type of laboratory. The sponsor submits the study reports to the FDA in food and color additive petitions, investigational new drug applications, new drug applications, new animal drug applications, biological product license applications, and other requests for permission to market a product.

Food and Drug Administration scientists evaluate the safety studies to determine whether or not the results support a conclusion that the product can be used safely. Until the mid-1970s, the underlying assumption in the agency review was that the reports submitted to the agency accurately described study conduct and precisely reported the study data. A suspicion that this assumption was mistaken was raised in the agency's review of studies submitted by a major pharmaceutical manufacturer in support of new drug applications for two important therapeutic products. Review scientists observed data inconsistencies and evidence of unacceptable laboratory practices in the study reports.

The FDA's Bureau of Drugs requested a "for-cause" inspection of the manufacturer's laboratories to determine the cause and extent of the discrepancies. A for-cause inspection is one initiated at the request of an agency unit when there is reason to suspect a problem in an FDA-regulated product. The authority to make for-cause inspections is a general one under the FFDCA, but one that had rarely been applied to animal laboratories.

In a statement in a Senate hearing on July 10, 1975, Dr. Alexander M. Schmidt, commissioner of food and drugs, reported the preliminary results of further agency investigations [1]. The findings indicated defects in design, conduct, and reporting of animal studies. For-cause inspections were conducted at several laboratories and revealed similar problems. The nature and extent of the findings in these inspections raised questions about the validity of studies being submitted to the agency.

The deficiencies observed in these inspections were summarized in the preamble to the proposed good laboratory practice regulations [2] as follows:

1. Experiments were poorly conceived, carelessly executed, or inaccurately analyzed or reported.
2. Technical personnel were unaware of the importance of protocol adherence, accurate observations, accurate administration of test substance, and accurate record keeping and record transcription.
3. Management did not assure critical review of data or proper supervision of personnel.
4. Studies were impaired by protocol designs that did not allow the evaluation of all available data.
5. Assurance could not be given for the scientific qualifications and adequate training of personnel involved in the research study.
6. There was a disregard for the need to observe proper laboratory, animal care, and data management procedures.

7. Sponsors failed to monitor adequately the studies performed in whole or in part by contract testing laboratories.
8. Firms failed to verify the accuracy and completeness of scientific data in reports of nonclinical laboratory studies in a systematic manner before submission to the FDA.

The problems were so severe in Industrial Bio-Test Laboratories (IBT) and Biometric Testing Inc. that both laboratories ceased doing preclinical studies. Industrial Bio-Trust Laboratories had been one of the largest testing laboratories in the United States, with thousands of its studies serving to support the safety of drugs, pesticides, and food additives. The FDA and the Environmental Protection Agency (EPA) began reviewing all the compounds that relied on IBT and Biometric Testing Inc. studies for support of safety. The agencies required the study sponsors to submit outside audits of the study data. From the audits of the IBT studies, EPA found 594 of 801 key studies, or 85% to be invalid [3]. The FDA's Bureau of Foods found 24 of 66 IBT studies, or 36% to be invalid [4].

Criminal charges of fraud were brought against four IBT officials. Three of the officials were convicted; a mistrial was declared in the case of the fourth official because of illness [5].

FDA'S RESPONSE TO THE PROBLEM

The conclusion that many studies on which the safety of regulated products had been based could be invalid was alarming to the FDA, the EPA, Congress, the public, and industry. Commissioner Schmidt established the Bioresearch Monitoring Program in early 1976 to develop a program that would deal with the problem of data validity not only in the area of safety studies but also in clinical testing. Congress voted a special appropriation of $16 million and additional personnel to support the program.

A steering committee, chaired by the associate commissioner for compliance and composed of the associate commissioners, the bureau directors, the chief counsel, the director of the National Center for Toxicological Research, and the executive director for regional operations, directed the program. Four task forces—the Toxicology Laboratory Monitoring Task Force, the Investigator Sponsor Task Force, the Institutional Review Committee Task Force, and the Administrative Task Force—handled different components of the program. The responsibility for developing a strategy to ensure the validity and reliability of all nonclinical laboratory studies to support the safety of FDA-regulated products was assigned to the Toxicology Monitoring Task Force. This task force was instructed to inventory all firms submitting research to the FDA and other involved federal agencies; to develop formal agreements with other agencies for the inspection of laboratories; to develop and publish standards for measuring the performance of research laboratories; to develop agencywide enforcement strategies; and to develop plans for hiring, training, and assigning the new employees authorized by Congress for the program.

The Toxicology Monitoring Task Force chose the publication of GLP regulations as the best approach for assuring study validity. Six other approaches were considered but were discounted as not feasible or efficient.

- One approach would have been to continue the program of for-cause inspections, but they would be triggered only by perceived deficiencies in the data after submission to the agency, and thus would not have provided systematic assurance that all studies were valid or guidance to laboratories on standards for conduct of studies.
- A second approach would have been to shift responsibility for nonclinical testing of regulated products to

the FDA. Such a shift would have required congressional authorization, because the FFDCA clearly places this responsibility on the sponsor of the product. In addition, the costs of such a shift would have been prohibitive.

- The third approach considered was for the agency to publish detailed test protocols and procedures for studies on regulated products. This, however, would have discouraged the use of informed scientific judgment in designing tests and inhibited the development of new toxicological methods.
- Another approach would have been to establish licensing procedures for testing laboratories, but developing uniform licensing criteria would have been very difficult, considering the variety of regulated products, test types, and laboratory facilities.
- Still another approach was the establishment of a full-time, on-site inspection program for laboratories similar to the U.S. Department of Agriculture's inspections of meat-processing plants. Such a program was considered to be an inefficient use of the FDA's investigational resources, because many testing facilities are too small or too diversified to justify full-time, on-site monitoring.
- Consideration was also given to the publication of GLP guidelines rather than regulations. While this would have provided the testing facilities with standards of conduct, it would not have given the agency an enforcement mechanism to ensure that the standards were met.

The regulations approach had several advantages. It was within the legal mandates of the agency and allowed efficient use of agency resources for ensuring compliance. It was also similar to the use of good manufacturing practice (GMP) regulations with which most of the regulated industries were already familiar. The main advantage, however, was that the regulations approach focused on the process by which testing facilities carried out studies rather than on the product being tested or the studies themselves. The use of scientific judgment in the planning and conduct of safety studies thus was not hampered, and the detail required for a focus on specific studies, or kinds of studies, was avoided.

Once the decision to establish GLP regulations had been made, a subcommittee was appointed to draft the regulations. This subcommittee was composed of individuals representing all the FDA bureaus and a variety of scientific disciplines. The subcommittee began its work with a rough draft that had already been prepared by personnel in the Bureau of Drugs. This early draft had used two independent, unsolicited sets of GLP guidelines submitted by G. D. Searle and Co. and the Pharmaceutical Manufacturers Association. The subcommittee's first draft was circulated to all FDA bureaus for comment, revised on the basis of these comments, and then circulated to other government agencies for comment. The subcommittee considered these comments in preparing the final draft, which was published as the proposed GLP regulations on November 19, 1976. The proposed regulations were designated as a new part 3.e. of Chapter 21 of the Code of Federal Regulations, but the final regulations were codified as part 58 (21 CFR Part 58).

FDA'S PROPOSED REGULATIONS

The purpose of the GLP regulations is to assure the quality and integrity of the data submitted to the FDA in support of the safety of regulated products. To this end, most of the requirements of the proposal would have been considered familiar and reasonable by any conscientious scientist. Protocols and standard operating procedures (SOPs), adequate facilities and equipment, full identification of test substances, proper animal care, equipment maintenance, accurate recording of observations, and accurate reporting of results are basic necessities for the conduct of a high-quality, valid toxicity, or any scientific study. The proposed regulations also placed a heavy emphasis on data recording and record and specimen retention to ensure that a study could be reconstructed at a later time if the need arose.

The proposed regulations went beyond these basic requirements for a valid study by requiring each study to have a study director who would have "ultimate responsibility for implementation of the protocol and conduct of the study" [§ 3e/31(a)], and each testing facility to have a quality assurance unit to monitor conduct of studies. The concept of a quality assurance unit to monitor study conduct was a new one to most laboratories but a familiar one in manufacturing facilities operating under various GMP regulations.

In addition, because the GLPs were regulations, the proposal identified the scope of the regulations, the authority under which they were promulgated, and the strategy for their enforcement.

Scope

The Toxicology Monitoring Task Force had not specified what types of studies would be considered to be within the scope of the GLPs. The subcommittee that drafted the regulations defined a nonclinical laboratory study as "any in vivo or in vitro experiment in which a test substance is studied prospectively in a test system under laboratory conditions to determine its safety" [§ 3e.3(d)]. The proposal explained that the term was to include only those studies conducted for submission to the FDA in support of an "application for a research or marketing permit." This latter term was a means of referring to the numerous categories of data required to be submitted to the agency, such as food and color additive petitions, new drug applications, and new animal drug applications. The studies covered by the regulations included all kinds of toxicity studies—from in vitro mutagenicity studies to acute, subchronic, and long-term toxicity/carcinogenicity studies—in which inadequate effectiveness might affect safety. Studies excluded from the scope of the regulations were those utilizing human subjects, clinical studies or field trials in animals, basic exploratory studies, or studies to determine physical or chemical properties of a test substance independent of a test system.

The proposal recognized that the scope might justifiably be defined on a different basis, possibly on a facilities basis, and asked for comments on whether specific types of testing facilities might be excluded from coverage by the regulations.

Enforcement Strategy

The basic mechanism of enforcement was to be inspection of testing facilities by FDA field investigators. The FDA's authority to conduct inspections of facilities engaged in interstate commerce of regulated products is well established, and such inspections are the primary method of enforcement of the FFDCA. Under the proposal, studies performed by a testing facility that refused to permit inspection would not be accepted in support of an application for a research or marketing permit.

At the conclusion of an inspection, the FDA investigator notifies the facility of any deficiencies identified during the inspection, both in writing (on Form 483, "Notice of Inspectional Observations") and in discussion with management. If the deficiencies were of a kind that might affect study validity, more formal warnings would be issued to the testing facility through a regulatory letter or a notice of adverse findings.

Initial planning under the Bioresearch Monitoring Program called for each testing facility to be inspected yearly. It was later decided that a biennial inspection would suffice to ensure that all 2-year studies would be inspected at least once while in progress.

When deficiencies were extensive enough to affect the validity of a study, the proposal provided that the study would not be considered by the FDA in support of a research or marketing permit. The proposal noted that the data from such a study had to be submitted to the agency, however, and that if they were adverse to the product might still be used as a basis for regulatory action. This difference in treatment was justified by the consideration that a bad study might reveal an adverse effect but could not establish the absence of an adverse effect.

The final and most severe enforcement strategy under the proposal was the disqualification of a testing facility. Data from a disqualified facility would not be accepted in support of a research or marketing permit. The agency viewed this penalty as one that would only be employed in cases in which the testing facility had severe, widespread deficiencies that raised questions about the validity of all the studies performed in the facility and in which previous regulatory efforts had failed to bring the facility into compliance with the regulations. Unlike the other enforcement strategies, there was no specific authority for disqualification; the GLP regulations themselves established this authority.

Authority

The GLP regulations were issued under the general mandate of section 701(a) of the FFDCA, which empowers the commissioner to promulgate regulations for the efficient enforcement of the act. The commissioner's power to issue regulations for determining that a clinical investigation of a drug intended for human use be scientifically reliable and valid [21 CFR 314.111(a)(5)] had been upheld by the Supreme Court in the decision *Weinberger* v. *Hynson, Westcott and Dunning, Inc.*, 412 U.S. 609 (1973). The clinical investigations regulations had also been issued under section 701(a) of the FFDCA. It was further considered that the authority to issue GLP regulations gave the agency the authority to establish the terms on which it would accept nonclinical testing data; therefore the proposed regulations provided for the rejection of studies if the testing facilities refused to permit inspection. The FDA already had the authority to compel inspection of nonclinical laboratories doing work on new drugs, new animal drugs, or medical devices. The FDA may inspect both manufacturing establishments and laboratories concerned with drugs and devices and examine research data on these products under section 704(a) of the FFDCA.

COMMENTS ON THE PROPOSAL AND THE FINAL REGULATIONS

More than 1000 individual items were contained in 22 oral responses from a 2-day public hearing and 174 written responses to the proposal. Many responses commented on both general issues, such as scope, and specific details in individual sections and paragraphs. The preamble to the final regulations addressed these comments in detail, and modifications, both substantial and editorial, were included in the final regulations, which were issued on December 22, 1978, and became effective June 20, 1979 [6].

Management and the Study Director

As outlined in the proposal, comments on the responsibilities of the study director identified many of these responsibilities as the prerogative of management. In response to these comments, a new section (§ 58.31) was included in the final regulations. This section established that if necessary, the management of the testing facility had the responsibility for designating and replacing the study director; for providing a quality assurance unit and assuring the actions to correct deviations reported by the quality assurance unit are taken; for assuring that the personnel and the tools (e.g., facilities and equipment) are available as needed; and for assuring that test and control articles are appropriately identified.

Despite making management responsible for many areas that the proposal had assigned to the study director, the final regulations retained the concept of the study director as the single focus of responsibility for study conduct by redefining the function of the study director as "overall responsibility for the technical conduct of the study, as well as for the interpretation, analysis, documentation and reporting of results, and represents the single point of study control" (§ 58.33).

The Quality Assurance Unit

Not surprisingly, many comments objected to the requirement for a quality assurance unit on the basis of increased costs, administrative burden, and interference with management prerogatives and informed scientific judgment of study directors. An alternative solution for study monitoring was not suggested, however.

The FDA retained the requirement for a quality assurance unit, or function, to monitor studies for conformance to the regulations. It was emphasized that the function was administrative rather than scientific. The personnel responsible for quality assurance for a given study were required to be separate from, and independent of, the personnel responsible for the direction and conduct of that study.

Many commentators wanted the inspection records compiled by the quality assurance unit excluded from the records to be inspected by the agency on the basis that an inspection "might violate the constitutional privilege against compelled self-incrimination." The agency rejected this argument, because the privilege against compelled self-incrimination is not available to a collective entity, such as a business enterprise, or to an individual acting as a representative of a collective entity. The agency did, however, exclude the quality assurance unit's inspection records form inspection to encourage more forthrightness in the reports. The quality assurance unit was required to certify that the inspection of studies and final reports had been made by means of a signed statement to be included in the final report [§ 58.35(b)(7)].

Scope

In general, the comments on the proposed regulations sought limitations through exclusion of various classes of FDA-regulated products, such as medical devices; various types of facilities, such as academic and not-for-profit organizations; or various types of studies, such as short-term studies. These suggestions were rejected primarily because the basic purpose of the regulations—to ensure the validity of safety data submitted to the agency—would have been frustrated by excluding particular products, facilities, or studies from coverage. None of the commentators suggested an alternative overall approach to defining the scope of the regulations.

The scope adopted in the final regulations was only slightly changed from the proposal; the main difference was the exclusion of functionality studies from coverage.

Inspections

The major concerns of the commentators with respect to the actual inspection of facilities were the competence and scientific qualifications of the FDA investigators. In early inspections (both the for-cause inspections prior to the proposal and the inspections made in the pilot program under the proposal), the agency assigned its most experienced field investigators and sent agency scientists to participate in the inspections. To further assure the competence of the investigators, a training program was established at the National Center for Toxicological Research for both field investigators and scientists. The compliance program for the GLPs also provides for scientific review in FDA headquarters of all GLP inspection reports.

That testing facilities still doubt the competence of some field investigators was evident in a comment on the 1987 revision of the GLPs [7], which requested training in the GLPs for the FDA's field personnel.

Disqualification

Numerous comments were made on the provisions for disqualification of a testing facility (subpart K). Although the proposal stated that the agency considered that it would only rarely invoke this penalty, it appeared from the objections that industry had interpreted these provisions to mean the agency would invoke disqualification frequently and for minor failures to comply with the regulations. On the basis of the objections, the sections of subpart K on purpose (§ 58.200) and the grounds for disqualification (§ 58.202) were extensively revised. The revision stated that the purposes of disqualification were as follows:

1. To permit the exclusion of completed studies from consideration in safety evaluation until it could be shown that noncompliance with the regulations did not affect the validity of the study data
2. To permit the exclusion of studies completed after disqualification from consideration in safety evaluation until the facility could demonstrate that it would conduct studies in compliance with the regulations.

Three grounds for disqualification were given in the final regulations; all three must be present to justify disqualification.

1. Failure of the facility to comply with one or more of the GLP regulations or other regulations applying to facilities published in Chapter 21 of the Code of Federal Regulations
2. Adverse affects on the validity of the studies
3. Failure to achieve compliance with regard to lesser regulatory actions, such as warnings or rejection of studies

EVALUATION OF THE FDA PROGRAM

The proposed GLP regulations announced that based on the requirements of the proposal, the FDA would conduct a number of surveillance inspections of testing facilities during November and December of 1976 and January of 1977. These inspections had the dual purpose of determining the status of the laboratories and evaluating the work ability of the proposed regulations. The results of this pilot inspection program were analyzed and published by the FDA's Office of Planning and Evaluation [8].

Forty-two laboratories were identified for inspection. Ongoing and completed studies would be examined as available. The inspections used a checklist that was divided into two parts, one part covering laboratory operations and the other study conduct. The checklist arbitrarily placed mixing and storing of test substances in the area of laboratory operations and distribution and characterization of the substances in study conduct.

In the completed survey, only 39 laboratories, with 67 studies, yielded usable data. Twenty-three of the testing facilities were sponsor laboratories, 11 contract laboratories, and five university laboratories. Forty-eight of the studies were completed and 19 ongoing. The findings showed that sponsor laboratories met 69% of the requirements, the contract laboratories met 56% of the requirements, and university laboratories met only 46% of the requirements.

Requirements in the areas of facilities, animal care, and personnel were the most often met, while the fewest requirements were met in the areas of the quality assurance unit, mixing and storage of the test substances, and record retention.

Ongoing studies showed better adherence (73% of the requirements met) than did completed studies (57%). Animal care and test substance distribution showed the greatest degree of adherence. Low degrees of adherence were found in the quality assurance function and protocol-related requirements. The comments of the agency investigators indicated that testing facilities were already making changes in their ongoing studies to bring them into compliance.

Following publication of the final regulations, a second survey was conducted to measure compliance against the final requirements [9]. The study sample consisted of 17 sponsor laboratories, 10 contract laboratories, and one university laboratory. The average compliance rate was 88%, with the deficiencies observed in sponsor and contract laboratories showing little difference. Compliance was measured both by the average compliance rate with the requirements of a section of the regulations or by the number of laboratories failing to meet one or more of the section's requirements. The following sections showed high compliance by both measurements: personnel, management, study director, general facilities, and facilities for animal care, handling of test and control articles, laboratory operations, specimen and data storage, record retention, and personnel and administration. Areas that showed low compliance by the same measures were quality assurance units, maintenance and calibration of equipment, SOPs, animal care (primarily the failure to analyze feed and water for interfering contaminants), test and control article characterization, mixtures of articles with carriers, study protocol, and study conduct (primarily failure to sign and date data sheets or to follow the protocol).

The results of these surveys indicated both the practicality of the regulations and the success of the vigorous efforts that most testing facilities were making to achieve compliance. The record of compliance continued to be good. In its 1984 update of compliance results [10], the FDA reported that 72% of the inspection reports since 1976 showed few or no substantial deviations from the regulations and 23% showed minor to significant deviations that could be corrected voluntarily by the testing facility. Four percent of the reports, however, showed significant deviations requiring corrective action within a specified period of time, and studies are still occasionally rejected because significant deviations render them invalid.

THE PROBLEM FROM EPA'S PERSPECTIVE

The EPA had concerns similar to those of the FDA. Under section 4 of the Toxic Substances Control Act (TSCA), the EPA evaluates laboratory data submitted to the agency regarding tests of the health effects of chemical substances and mixtures. Also, under authority of the Federal Insecticide, Fungicide and Rodenticide Act (FIFRA), the EPA evaluates laboratory test data relating to hazards to humans arising from the use of a pesticide product when the agency evaluates pesticide registration applications.

The EPA was aware of the problems the FDA had uncovered in the mid-1970s relating to unacceptable laboratory practices. The EPA responded to the FDA's findings by forming the toxicology auditing program in the agency's Office of Pesticide Programs. The EPA also held public hearings to solicit comments on how appropriate the agency's approach was to data quality assurance for pesticide testing.

In 1978, the EPA and FDA formalized both agencies' commitment to establish a coordinated quality assurance program through an interagency agreement. Under this agreement, the FDA provided assistance during EPA data audits. Between 1978 and 1979, the agencies performed 65 joint audits that indicated that some testing facilities did not follow GLPs. The EPA referred some of these facilities to the Department of Justice for prosecution.

EPA'S PROPOSED REGULATIONS

Like the FDA, the EPA considered different approaches to assure that data submitted to the agency complied with necessary quality standards.

Licensing or certification of laboratories was considered impractical for toxicology laboratories because of the great diversity and range of testing capabilities and the complex quality control procedures used in toxicology testing.

A voluntary standard-setting scheme administered by the private sector was rejected because such schemes were considered practically unenforceable.

Like the FDA, the EPA determined that the promulgation of GLP regulations would most effectively handle the problem of compliance with adequate control standards, and the agency published proposed health effects standards for testing under TSCA on May 9, 1979 [11]. Proposed GLP regulations applicable to laboratory studies submitted to the EPA in compliance with FIFRA were published on April 18, 1980 [12]. Supplemental GLP standards for the development of data on physical, chemical, persistence, and ecological effects of chemical substances for which the EPA requires testing under section 4 of TSCA were published on November 21, 1980 [13]. The EPA took this action because the previously published GLPs for health effects testing did not address the analytical problems associated with physical, chemical, and persistence testing.

Differences Between EPA Proposed Regulations and FDA Regulations

When it issued proposed GLP regulations in 1978 and 1980, the EPA harmonized those regulations that the final GLP regulations which had been issued by the FDA in 1978. There were major differences, however, because the two agencies' approaches to regulating laboratory studies differed. The specific working of various sections of the EPA's proposed regulations varied from those of the FDA because of the differing scope of the authority of each agency.

EPA's Final Regulations

The EPA's FIFRA and TSCA GLP regulations were both issued in final form on November 29, 1983 [14]. The FIFRA GLP regulations were codified as 40 CFR 160, and the TSCA GLP regulations as 40 CFR 792. In terms of the TSCA GLPs, the final regulations incorporated the proposed GLPs issued on May 9, 1979 and November 21, 1980.

GLP REVISIONS IN THE 1980S

FDA Revisions

In 1984, the FDA proposed revising its 1978 GLP regulations. The rationale for this revision was to clarify, amend, or delete provisions of the regulations in order to reduce the regulatory burden on testing facilities.

During agency inspections, the FDA had found that most laboratories were complying with the GLP requirements—indeed, that the violations it had noted in the mid-1970s were the exception, rather than the general rule—and the agency thought that it could streamline the regulations without compromising the GLP program. The FDA had also received comments and questions about the GLP regulations that indicated that several GLP provisions did not significantly contribute to the quality and integrity of data submitted to the agency. At the same time, the agency was undertaking a review of its regulations to minimize regulatory burdens.

The FDA established a GLP review task team to identify provisions in the regulations that could be amended or deleted, and this team recommended revisions to 36 GLP provisions. Recommendations were issued as a proposed rule on October 29, 1984 [15]. The proposal made various changes to definitions to reduce the amount of paperwork required for nonclinical laboratory studies and to clarify earlier GLP provisions. Similar clarifications were made to the provisions, delineating the definition and function of the study director and quality assurance unit.

In the 1984 proposal, changes were also made to inform collection requirements subject to the Paperwork Reduction Act of 1980. Modifications were made to the provisions regarding animal care, animal supply, and administrative and personnel facilities. Provisions regarding equipment design, maintenance and calibration of equipment, SOPs, animal care, test and control article characterizations, and mixtures of articles with carriers were changed to allow more flexibility of laboratory operations. The section on laboratory protocols was amended to eliminate unnecessary entries by allowing laboratories to identify the information applicable to the articles being tested. The agency also deleted the requirements that the selection of the test system be justified in the protocol. Other changes to the GLP regulations involved revisions to provisions regulating conduct of laboratory studies and the storage, retrieval, and retention of records.

The FDA received 33 comments on its proposed GLP revisions. After considering these comments, the agency issued its final GLP provisions on September 4, 1987 [16]. Some of the comments received by the agency indicated a need to add new terms to the definition section of the regulations (e.g., *study initiation* and *study completion*), while others encouraged the FDA to retain the original GLP language in certain provisions rather than make the amendments the agency had proposed in 1984.

EPA Revisions

Among the comments received by the FDA, eight comments urged the agency to encourage the EPA to adopt similar revisions to its GLP regulations, which were now more stringent than the FDA's regulations. The FDA stated that the agency consulted with the EPA regarding the changes made to the FDA's regulations, and that the FDA would cooperate with the EPA when the latter agency revised its own GLP regulations. As a result of its own monitoring of GLP compliance, EPA agreed that its own GLP regulations could be streamlined without compromising the integrity of data submitted to the agency.

The EPA's proposed revisions to its FIFRA and TSCA GLP regulations were issued on December 28, 1987 [17]. The EPA agreed with the FDA that many GLP provisions could be amended to incorporate the changes that had been made by the FDA. In addition, the scope of the FIFRA regulations was expanded to include environmental testing provisions that already existed in the TSCA GLPs, and to include product performance data (efficacy testing). The EPA also proposed changes. Some changes were made to the proposed regulations in response to these comments, such as exempting from routine EPA inspections the records of quality assurance unit findings and problems, as well as records of corrective actions recommenced and taken, except under special circumstances. The final versions of EPA's revisions to its GLPs were issued on August 18, 1989 [18].

The EPA's proposed GLP revisions basically conformed to the charges the FDA had made in the latter agency's final rule of September 4, 1987. The major differences between the EPA proposals and the FDA GLPs continued to reflect the varying needs and responsibilities of each agency and the expanded scope of the EPA's regulations in light of the testing and test systems affected under the EPA's authority to require test data in support of research or marketing permits to include ecological effects, environmental and chemical fate, and efficacy testing in addition to health effects testing.

Other federal agencies, as well as international agencies and organizations, also developed GLP programs. The National Toxicology Program concluded that studies performed under contract to the program should be performed in compliance with GLPs, and established a quality assurance function to monitor the laboratories and studies. In 1981, the Organization for Economic Cooperation and Development (OECD) developed GLP principles for studies performed for the European Economic Community (EC) countries. Between 1986 and 1988, EC council directives adopted the OECD and required that all EC countries monitor and verify compliance with those standards.

In 1982, the Japanese Ministry of Health and Welfare issued GLP standards for safety studies on drugs. This was followed in 1984 by GLP standards issued for studies on industrial chemicals by the Japanese Ministry of International Trade and Industry and GLP standards for toxicological studies on agricultural chemicals by the Japanese Ministry of Agriculture, Forestry, and Fisheries. There are differences in these regulations and guidelines that pose problems for sponsors planning studies to meet the requirements of different agencies or countries [19].

As a solution to part of this problem, the FDA has developed memoranda of understanding (MOUs) with Canada

(1979), Sweden (1979), Switzerland (1985), France (1986), Italy (1988), Germany (1988), the Netherlands (1988), and the United Kingdom (1988). These MOUs acknowledge mutual recognition of the adequacy of inspectional programs in the participating countries and permit the exchange of data between the countries without need for independent verification by the recipient country.

REFERENCES

1. Statement by Alexander M. Schmidt, M.D., commissioner, Food and Drug Administration, Public Health Service, Department of Health, Education and Welfare before the Subcommittee on Health, Committee on Labor and Public Welfare and Subcommittee on Administrative Practice and Procedure, Committee on the Judiciary, U.S. Senate, July 10, 1975.
2. Department of Health, Education and Welfare, Food and Drug Administration. Nonclinical laboratory studies: Proposed regulations for good laboratory practice. *Fed Reg* 41:51205–51230, 1976.
3. Environmental Protection Agency, Office of Pesticide Programs. Summary of the IBT Review Program. July 1983.
4. Survey by Dr. Jean M. Taylor.
5. IBT data falsification. *Food Chem News* 25(24):2, 1983.
6. Department of Health, Education and Welfare, Food and Drug Administration. Good laboratory practice regulations. *Fed Reg* 43:59985–60024, 1978.
7. Department of Health, Education and Welfare, Food and Drug Administration. Good laboratory practice regulations *Fed Reg* 52:33767–33782, 1987.
8. Food and Drug Administration, Office of Planning and Evaluation. OPE Study 42: Results of the Nonclinical Laboratory Good Laboratory Practices Pilot Compliance Program. Sept. 1977.

9. Food and Drug Administration, Office of Planning and Evaluation. OPE Study 49: Results of the Toxicology Laboratory Inspection Program (Jan.–March 1979). July 1979.
10. Department of Health, Education and Welfare, Food and Drug Administration. 21 CFR Part 58: Good laboratory practice regulations: Proposed rule. *Fed Reg* 49: 43529–43537, 1984.
11. Environmental Protection Agency. Good laboratory practice standards for health effects: Proposed rule. *Fed Reg* 44: 27362–27375, 1979.
12. Environmental Protection Agency. Pesticide programs, Guidelines for registering pesticides in the United States: Proposed good laboratory practice guidelines of toxicology testing, proposed rule. *Fed Reg* 45:26373–26385, 1980.
13. Environmental Protection Agency. Physical, chemical, persistence, and ecological effects testing: Good laboratory practice standards, proposed rule. *Fed Reg* 45:77353–77365, 1980.
14. Environmental Protection Agency. Toxic substances control: Good laboratory practice standards; Final rule. *Fed Reg* 48: 53921–53944, 1983: Environmental Protection Agency. Pesticides program; Good laboratory practice standards: Final rule. *Fed Reg* 48:53945–53969, 1983.
15. Food and Drug Administration. Good laboratory practice regulations: Proposed rule. *Fed Reg* 49:43530–43537, 1984.
16. Food and Drug Administration. Good laboratory practice regulations: Final rule. *Fed Reg* 52:33768–33782, 1987.
17. Environmental Protection Agency. Federal Insecticide, Fungicide and Rodenticide Act (FIFRA): Good laboratory practice standards, proposed rule. *Fed Reg* 52:48920–48933, 1987; Environmental Protection Agency. Toxic Substances Control. Act (TSCA): Good laboratory practice standards, proposed rule. *Fed Reg* 52: 489933–48946, 1987.
18. Environmental Protection Agency. Toxic Substances Control Act (TSCA): Good laboratory practice standards, final rule. *Fed Reg* 54: 34034–34050, 1989; Environmental Protection Agency. Federal Insecticide, Fungicide and Rodenticide Act

(FIFRA): Good laboratory practice standards, final rule. *Fed Reg* 54:34052–34074, 1989.

19. For differences in worldwide GLP standards, see Chapter 3 of this book and also RS DeWoskin, SM Taulbee. *International GLPs*. Interpharm Press, 1993.

2

FDA/GLP Regulations

WENDELL A. PETERSON
Parke-Davis Pharmaceutical Research Division
of Warner-Lambert Company, Ann Arbor,
Michigan, U.S.A.

INTRODUCTION

Proposed good laboratory practice (GLP) regulations were published in 1976 [1]. Final regulations were published in 1978 [2]. The regulations were revised in 1980 [3] and 1987 [4], twice in 1989 [5,6], and again in 1991 [7].

Good Laboratory Practice regulations [8] are promulgated by the commissioner of the U.S. Food and Drug Administration (FDA) under general authority granted by section 701(a) of the Federal Food, Drug, and Cosmetic (FD&C) Act [9]. Unlike current good manufacturing practice (CGMP) regulations [10], which can be referenced back to specific statutory language [the words current good manufacturing practice in section 501(a)(2)(B)], the term “good laboratory practice” does not appear in the FD&C Act; rather, the GLP regulations are issued under the FDA commissioner’s implied powers to prescribe standards for the conduct of studies designed to establish the safety of products regulated by the FDA.

This chapter provides a general discussion of all aspects of FDA’s GLP regulations, as amended to September 13, 1991. Where appropriate, FDA interpretations are presented for specific sections of the GLP regulations. For critical parts of the regulations, a more in-depth discussion is provided, including means for implementation and an evaluation of positive and negative impacts on the conduct of GLP-regulated studies.

SUBPART A: GENERAL PROVISIONS

§ 58.1: Scope

(a) This part prescribes good laboratory practices for conducting nonclinical laboratory studies that support or are intended to support applications for research or marketing permits for products regulated by the Food and Drug Administration, including food and color additives, animal food additives, human and animal drugs, medical devices for human use, biological products, and electronic products. Compliance with this part is intended to assure the quality and integrity of the safety data filed pursuant to sections 406, 408, 409, 502, 503, 505, 506, 507, 510, 512–516, 518–520, 706, and 801 of the Federal Food, Drug, and Cosmetic Act and sections 351 and 354–360F of the Public Health Service Act.

(b) References in this part to regulatory sections of the Code of Federal Regulations are to Chapter I of Title 21, unless otherwise noted.

This preamble to the GLP regulations [2], the report of a series of three briefing sessions on the GLP regulations that were conducted by FDA on May 1, 2, and 3, 1979 [11], and the collections of responses by Dr. Paul Lepore (FDA spokesman on GLP issues) to questions about the GLP regulations [12], have defined the types of studies to which the GLP regulations apply. In general, *all* of the following conditions must exist before a study will be regulated by GLP:

1. Study of a product regulated by the FDA (except cosmetics).
2. In vivo or in vitro study.
3. Study in which the FDA-regulated product is administered or added to nonhuman animals, plants, micro-organisms, or subparts of the preceding.
4. Study results submitted or intended to be submitted to FDA in support of (i.e., as the basis for) the approval of an application for a research or marketing permit.
5. Study results may be used to predict adverse effects of and/or to establish safe use characteristics for the FDA-regulated product.

The FDA has made it clear that the duration of the study and the place where the study is conducted do not determine whether or not the study is GLP-regulated. Thus, the GLPs apply to short-term studies (e.g., median lethal dose studies and irritation studies) as well as long-term studies that meet all of the above criteria, and the GLPs apply to such studies whether conducted in a manufacturer's laboratories, in a university laboratory, or at a contract or subcontract facility. The FDA expects GLP compliance for studies conducted in foreign countries as well as for those conducted within the United States.

Without attempting to provide a comprehensive listing, the following are examples of studies to which the GLPs can apply:

Ames test
E. Coli mutagenicity

Sister chromatid exchange
Bone marrow cytogenetic
In vivo cytogenetic
In vitro mutation
In vivo micronucleus
Chromosomal aberration
Median lethal dose (LD_{50})
Acute dermal toxicity
28-day dermal toxicity
Dermal irritation
Eye irritation
Venous irritation
Muscle irritation
Intra-arterial tolerance
Guinea pig maximization
Phototoxicity
Ototoxicity
Dependency tests on known or suspected addictive drugs
Target animal absorption, distribution, metabolism, and excretion (ADME)
Subchronic (up to 13 weeks' duration, multiple dosing, any route of administration)
Chronic (6 months or longer in duration, multiple dosing, any route of administration)
Study of fertility in early embroyonic development (previously referenced as segment I)
Perinatal/postnatal (formerly referred to as segment III)

To reiterate, the foregoing list is only intended to illustrate the wide range of studies that may be GLP-regulated.

Examples of studies that are not within the scope of the GLP regulations include the following:

Pharmacology experiments
Basic research
Dose-range finding studies
Studies to develop new methodologies

Human or animal efficacy studies
Chemical assays for quality control of commercial products
Stability tests on finished dosage forms and products
Tests for conformance to pharmacopeial standards
Exploratory studies on viruses and cell biology
Tests of functionality and/or appropriateness of food additives
Tests of extract ability of polymeric materials that contact food
Chemical tests used to derive the specifications of marketed food products
Studies on medical devices that do not come in contact with or are not implanted in humans
Tests of diagnostic products
Chemical and physical tests of radiation products
Tests conducted for the release of licensed biological products

The foregoing list is also intended to be illustrative, and not comprehensive.

A facility that conducts both GLP-regulated and non-GLP-regulated studies should think carefully about attempting to maintain a dual standard in any one laboratory or with any one group of laboratory workers. In the author's experience, such a dual standard is very difficult to maintain without carryover of non-GLP standards to GLP-regulated work. In such a case it may be far better to maintain a general GLP standard (e.g., data collection, record keeping) for all work in the laboratory, but perhaps allow exceptions for the non-GLP studies in areas such as quality assurance (QA) inspections and analytical requirements for test and control articles and article/carrier mixtures.

The effective date of the GLP regulations was June 20, 1979. The regulations did not apply retroactively, therefore studies begun and completed prior to the effective date were not required to comply with the GLPs even if submitted to

FDA on or after June 20, 1979. For studies in progress on June 20, 1979, only those portions of the study done on or after June 20 were required to be done in compliance with the regulations. Of course, those studies initiated on or after the effective date were to be done in full compliance with the GLPs.

§ 58.3: Definitions

A good understanding of the definitions in section 58.3 is critical to an interpretation of many of the other sections of the regulations.

An illustration of the importance of a definition to regulatory interpretation can be found in Environmental Protection Agency (EPA) regulations issued under the Resource Conservation and Recovery Act (RCRA). Among other things, these regulations are designed to regulate the disposal of "solid waste." Anyone relying on the normal definition of "solid" in interpreting RCRA requirements would make a grave error, because solid is defined in RCRA to include solid, liquid, semisolid, and contained gaseous materials.

Although there is nothing in the definitions section of the GLP regulations to rival RCRA's rewriting of the basic laws of chemistry and physics, a clear understanding of GLP definitions is essential to a proper interpretation of GLP requirements.

As used in this part, the following terms shall have the meanings specified:

(a) "Act" means the Federal Food, Drug, and Cosmetic Act, as amended (secs. 201–902, 52 Stat. 1040 et seq., as amended (21 U.S.C. 321–392)).

(b) "Test Article" means any food additive, color additive, drug, biological product, electronic product, medical device for human use, or any other article subject to regulation under the act or under sections 351 and 354–360F of the Public Health Service Act.

Note the wide range of FDA-regulated products to which the GLPs apply.

(c) "Control article" means any food additive, color additive, drug, biological product, electronic product, medical device for human use, or any article other than a test article, feed, or water that is administered to the test system in the course of a nonclinical laboratory study for the purpose of establishing a basis for comparison with the test article.

The term control article refers to materials that are administered or added to the typical control group that is part of most safety studies. The term includes materials commonly referred to as "positive controls" (e.g., a marketed drug that is administered or added to a positive control group as part of a study of an investigational drug of the same therapeutic category) as well as vehicles, solvents, and other carrier materials (other than water and animal diets) when such materials are given to control groups.

Although the GLP revisions of 1987 excluded animal feed and water from the definition of control article, it would appear that such common vehicles as saline solutions and carboxymethylcellulose solutions still fall within the definition. Such a strict definition of the term for such innocuous vehicles as saline solutions is quite burdensome when one considers the requirements for control articles that are found in other sections of the GLPs: characterization [§ 58.105(a)], stability testing [§58.105(b)], sample retention [§ 58.105(d)], and inventory [§ 58.107(d)]. It does not appear that this comprehensive definition is enforced by FDA field investigators in the course of GLP inspections.

Positive controls (usually known mutagens) used in mutagenicity studies also fall outside the definition of control article because they are administered to control groups for the purpose of establishing the ability of the assay to detect mutagenic activity and not for the purpose of "establishing a basis for comparison with the test article."

(d) "Nonclinical laboratory study" means in vivo or in vitro experiments in which test articles are studied prospectively in test systems under laboratory conditions to de-

> termine their safety. The term does not include studies utilizing human subjects or clinical studies or field trials in animals. The term does not include basic exploratory studies carried out to determine whether a test article has any potential utility or to determine physical or chemical characteristics of a test article.

Many of the issues relating to the definition of nonclinical laboratory study were addressed in the discussion of GLP § 58.1 (Scope). "Field trials in animals" includes all efficacy studies of new animal drugs. Such studies are outside the scope of the GLP regulations. This is consistent with the GLP exemption for human clinical trials. The exemption for "basic exploratory studies carried out to determine whether a test article has any potential utility" would extend to early screening studies of a test article, the results of which are used to determine whether a test article merits further development or not.

Good Laboratory Practice § 58.105(a) requires that all test articles be appropriately characterized. Compliance requires documentation that characterization has been done. The tests conducted to provide this documentation, however, are not GLP-regulated, although such tests will in many instances be subject to CGMP standards (e.g., when the test article will also be used in human clinical studies).

The GLP revisions of 1987 modified slightly the definition of nonclinical laboratory study by changing a few nouns, verbs, and adjectives from singular to plural. This now permits the conduct of several experiments using the same test article under a single comprehensive protocol or the concurrent test of several test articles using a single common procedure under a single protocol.

> (e) "Application for research or marketing permit" includes:
> (1) A color additive petition, described in part 71.
> (2) A food additive petition, described in parts 171 and 571.
> (3) Data and information regarding a substance sub-

mitted as part of the procedures for establishing that a substance is generally recognized as safe for use, which use results or may reasonably be expected to result, directly or indirectly, in its becoming a component or otherwise affecting the characteristics of any food, described in §§ 170.35 and 570.35.

(4) Data and information regarding a food additive submitted as part of the procedures regarding food additives permitted to be used on an interim basis pending additional study, described in § 180.1.

(5) An "investigational new drug application," described in part 312 of this chapter.

(6) A "new drug application," described in part 314.

(7) Data and information regarding an over-the-counter drug for human use, submitted as part of the procedures for classifying such drugs as generally recognized as safe and effective and not misbranded. Described in part 330.

(8) Data and information about a substance submitted as part of the procedures for establishing a tolerance for unavoidable contaminants in food and food-packaging materials, described in parts 109 and 509.

(9) Data and information regarding an antibiotic drug submitted as part of the procedures for issuing, amending, or repealing regulations for such drugs, described in § 314.300 of this chapter.

(10) A "Notice of Claimed Investigational Exemption for a New Animal Drug," described in part 511.

(11) A "new animal drug application," described in part 514.

(12) (Reserved)

(13) An "application for a biological product license," described in part 601.

(14) An "application for an investigational device exemption," described in part 812.

(15) An "application for Premarket Approval of a Medical Device," described in section 515 of the act.

(16) A "Product Development Protocol for a Medical Device," described in section 515 of the act.
(17) Data and information regarding a medical device submitted as part of the procedures for classifying such devices, described in part 860.
(18) Data and information regarding a medical device submitted as part of the procedures for establishing, amending, or repealing a performance standard for such devices, described in part 861.
(19) Data and information regarding an electronic product submitted as part of the procedures for obtaining an exemption from notification of a radiation safety defect or failure of compliance of a radiation safety defect or failure of compliance with a radiation safety performance standard, described in subpart D of part 1003.
(20) Data and information regarding an electronic product submitted as part of the procedures for establishing, amending, or repealing a standard for such product, described in section 358 of the Public Health Service Act.
(21) Data and information regarding an electronic product submitted as part of the procedures for obtaining a variance from any electronic product performance standard as described in § 1010.4.
(22) Data and information regarding an electronic product submitted as part of the procedures for granting, amending, or extending an exemption from any electronic product performance standard, as described in § 1010.5.

This section of the GLPs describes the various types of submissions to FDA that include safety information derived from studies that must be conducted in accordance with the GLP regulations.

(f) "Sponsor" means:
(1) A person who initiates and supports, by provision of financial or other resources, a nonclinical laboratory study;
(2) A person who submits a nonclinical study to the Food

and Drug Administration in support of an application for a research or marketing permit; or

(3) A testing facility, if it both initiates and actually conducts the study.

The definition of sponsor indicates who bears ultimate responsibility for a nonclinical laboratory study. A sponsor may assign the job of actual study conduct and/or reporting, but ultimate responsibility for the study cannot be delegated. The sponsor must thus assure that a nonclinical laboratory study is conducted in compliance with GLP standards, and must supply the statement of GLP compliance or description of GLP noncompliance (conforming amendments statement) that must accompany the submission to FDA of the results of a nonclinical laboratory study (Section XI). The definition does not preclude joint sponsorship of a study.

(g) "Testing facility" means a person who actually conducts a nonclinical laboratory study, i.e., actually uses the test article in a test system. "Testing facility" includes any establishment required to register under section 510 of the act that conducts nonclinical laboratory studies and any consulting laboratory described in section 704 of the act that conducts such studies. "Testing facility" encompasses only those operational units that are being or have been used to conduct nonclinical laboratory studies.

If a facility conducts nonclinical laboratory studies, it is a "testing facility" and is subject to inspection by FDA to determine its GLP compliance status. If a facility conducts nonclinical laboratory studies as well as studies that do not meet the definition of nonclinical laboratory study, then only those portions of the facility that conduct nonclinical laboratory studies are subject to a GLP inspection by FDA. The portions of the facility that conduct studies other than nonclinical laboratory studies are not subject to inspection by FDA unless FDA has inspectional authority under some other set of regulations.

(h) "Person" includes an individual, partnership, corporation, association, scientific or academic establishment, government agency, or organizational unit thereof, and any other legal entity.

This all-encompassing definition of person precludes the exemption of any person or legal entity from the definition of sponsor or testing facility if that person or other legal entity meets the definitions of those two terms.

(i) "Test system" means any animal, plant, microorganism, or subparts thereof to which the test or control article is administered or added for study. "Test system" also includes appropriate groups or components of the system not treated with the test or control articles.

In most instances the test system will be self-evident (e.g., the animal to which the test article is administered or applied). Studies with micro-organisms, however, sometimes present difficulty in defining the test system. In the case of the Ames test, for example, the test system is not merely the colonies of salmonella or yeast, but includes in addition the culture medium, metabolic activation agent (if any), biotin, histidine, and buffer (if any). The last sentence of the definition makes it clear that untreated control groups also meet the definition of test system even though a test or control article is not administered or applied to such groups.

(j) "Specimen" means any material derived from a test system for examination or analysis.

In most instances the specimens will be self-evident (e.g., samples of blood, plasma, serum, urine, spinal fluid, aqueous humor, organs, tissues, and tissue fractions that are taken from a test system with the intention of performing an examination or analysis). In other instances the definition may not be as clear. For example, the assay plates used in the mammalian cell transformation assay and the mammalian point mutation assay are considered specimens even though they bear many of the attributes of a test system. For these assays, the originally plated cells plus media and excipients are the test system. After treatment with the test or control article, however, the palates are stained and transformed cells are enumerated. The plates then become "material derived from the test system for examination or analysis"; in other words, specimens.

Care should be taken to distinguish specimen from "raw data," since GLP requirements differ for each. For example, it is often erroneously stated that a microscopic slide is raw data when in fact it is a specimen.

> (k) "Raw data" means any laboratory worksheets, records, memoranda, notes, or exact copies thereof, that are the result of original observations and activities of a nonclinical laboratory study and are necessary for the reconstruction and evaluation of the report of that study. In the event that exact transcripts of raw data have been prepared (e.g., tapes that have been transcribed verbatim, dated, and verified accurate by signature), the exact copy or exact transcript may be substituted for the original source as raw data. "Raw Data" may include photographs, microfilm or microfiche copies, computer printouts, magnetic media, including dictated observations, and recorded data from automated instruments.

Examples of raw data include records of animal receipt, records of animal quarantine, results of environmental monitoring, instrument calibration records, original recordings of such parameters as animal body weights or food consumption values, handwritten transcriptions to paper records of information displayed as a digital read-out on automated equipment, high-performance liquid chromatography (HPLC) tracings, integrator output from HPLC equipment, recorded clinical observations, a photograph of a lesion noted at autopsy, a pathologist's written or tape-recorded diagnosis of a microscopic slide, printed paper tapes containing recorded diagnosis of a microscopic slide, printed paper tapes containing values generated by hematology and blood chemistry equipment, values generated by hematology and blood chemistry equipment, and electrocardiographic tracings. These are only examples; the reader could expand the list 10- or 100-fold.

Microfilm and microfiche copies, carbon copies, or photocopies of original raw data may be substituted for the original raw data as long as they are exact and legible copies.

Cage cards that contain information such as animal number, study number, and treatment group are not raw data as long as no original observations are recorded on the card, nor are transformations of raw data (e.g., calculations of mean and standard deviation or other statistical values) considered raw data, because they can always be recalculated from the original raw data.

In the case of handwritten raw data, the original recording of information on paper constitutes the raw data that must be retained under § 58.190(a) of the regulations. Any subsequent transcriptions of this information will not substitute for the originally recorded information. Scientists and technicians will sometimes record raw data on scraps of paper or even on paper towels. Their intention is to neatly transcribe the information to official data forms at a later time and to discard the originally recorded data. This practice is to be discouraged, because the scraps of paper or paper towels are the real raw data and must be retained.

The FDA has indicated that a pathologist's interim microscopic diagnoses are not raw data because such diagnoses are not "necessary for the reconstruction and evaluation of the report of (a) study." Only when the pathologist signs off on a final diagnosis does that diagnosis become raw data.

The provision in the definition of raw data for the substitution of exact transcripts of raw data for the original has been narrowly construed by FDA. It applies only to the verbatim transcription of tape-recorded information (e.g., a pathologist's voice recording of a microscopic diagnosis or veterinarian's voice recording of a clinical observation) that is dated and verified as accurate by signature. In this case the original tape recording need not be retained.

If raw data are transcribed to a computer database, neither the electronically stored data nor the paper printout can substitute for the original. Information entered into the computer by direct data capture offers two options, however. The laboratory may elect to treat the electronically recorded information or a hard copy printout of the information as raw data. If the hard copy is retained, the magnetic media can be discarded or reused. If a laboratory elects to treat the magnetic media as raw data, it must retain an ability to display the data in readable form for the entire period during which that information is required to be retained. (See § 58.195 for a definition of required retention periods.) If a change in computer systems would entail the loss of the ability to display electronically stored data, the laboratory should generate hard copies of the data before the computer systems are changed.

> (l) "Quality assurance unit" means any person or organizational element, except the study director, designated by testing facility management to perform the duties relating to quality assurance of nonclinical laboratory studies.

Note the language "any person . . . except the study director." When read in conjunction with GLP § 58.35(a), it is clear that the person or persons designated to perform QA functions need not be full-time QA personnel. This flexibility is provided primarily to accommodate smaller laboratories in which the volume of GLP-regulated work is not sufficient to justify a full-time QA person. A person from the pharmacology department, for example, can perform the QA function for toxicology studies on a part-time basis, but spend the rest of his or her time in the conduct of pharmacology studies. Where the volume of work is sufficient to justify employing one or more full-time QA professionals, that is the preferred arrangement. Such an arrangement provides the degree of independence that is so important to the success of any quality program, removes the possibility that the demands of the part-time QA person's other responsibilities will interfere with his or her performance of the QA function, and allows more time for the development of expert audit and inspection skills. The author is aware of no major testing facility in the

United States in which the QA professionals are part-time, although there are instances in which the full-time QA staff is supplemented by temporary assignments from other departments.

Issues relating to the "quality assurance unit" (QAU) will be addressed in greater depth in the later discussion of GLP § 58.35.

(m) "Study director" means the individual responsible for the overall conduct of a nonclinical laboratory study.

Note the words *the individual.* There may be only one designated study director for any one study at any one time. It is not permissible, for example, to appoint a study director and an assistant study director, but it is permissible to name an alternate study director who will serve as study director only in the absence of the study director. It is FDA's intent that the study director serve as the single point of study control.

For a detailed description of the study director role, see the later discussion for GLP § 58.22.

(n) "Batch" means a specific quantity or lot of a test or control article that has been characterized according to § 58.105(a).

The GLP definition of batch differs from that found in FDA's CGMP regulations [§ 210.3(b)(2)] [10]. The CGMP definition relates any one batch to a defined cycle of manufacture. The GLP definition, on the other hand, relates batch to a characterization process; thus, for example, a GLP batch may be part of a CGMP batch or may be the result of a combination of two or more CGMP batches. The only GLP requirement is that a batch be characterized as to identity, strength, purity, and composition or other appropriate characteristics.

(o) "Study initiation date" means the date the protocol is signed by the study director.
(p) "Study completion date" means the date the final report is signed by the study director.

§ 58.10: Applicability to Study Performed Under Grants and Contracts

> When a sponsor conducting a nonclinical laboratory study intended to be submitted to or reviewed by the Food and Drug Administration utilizes the services of a consulting laboratory, contractor, or grantee to perform an analysis or other service, it shall notify the consulting laboratory, contractor, or grantee that the service is part of a nonclinical laboratory study that must be conducted in compliance with the provisions of this part.

The notification required by this section should be in writing. The form of the writing is not important from a GLP standpoint, but it may be advantageous to put the notification into a legally binding document (e.g., contract). Alternatively, the notification may appear, for example, in a study protocol signed by the sponsor or in a letter from the sponsor to the contractor.

High-volume contract laboratories often perform both GLP-regulated and non-GLP-regulated studies, so it is important to specify if a study is to be conducted under GLP conditions.

Some contract laboratories and professional consultants (e.g., veterinary ophthalmologists and pathologists), may not be familiar with the GLP regulations. In such cases, mere notification of a requirement to provide GLP-complying services may not be sufficient. It is advisable to spend time with contractors and professional consultants to review in detail the GLP requirements that will apply to the work they will perform. It is especially important to review with them the GLP requirements for documentation and document retention.

§ 58.15 Inspection of a Testing Facility

> (a) A testing facility shall permit an authorized employee of the Food and Drug Administration, at reasonable times and in a reasonable manner. To inspect the facility and

to inspect (and in the case of records also to copy) all records and specimens required to be maintained regarding studies within the scope of this part. The records inspection and copying requirements shall not apply to quality assurance unit records of findings and problems, or to actions recommended and taken.

(b) The Food and Drug Administration will not consider a nonclinical laboratory study in support of an application for a research or marketing permit if the testing facility refuses to permit inspection. The determination that a nonclinical laboratory study will not be considered in support of an application for a research or marketing permit does not, however, relieve the applicant for such a permit of any obligation under any applicable statue or regulation to submit the results of the study to the Food and Drug Administration.

All laboratories operating within the United States that conduct nonclinical laboratory studies are subject to inspection by FDA. Such inspections may include an inspection of laboratory facilities, laboratory records, and specimens. The FDA, however, has no legal authority to conduct such inspections outside the United States. Such inspections do occur, but only after a request from the FDA to conduct such an inspection has received the consent of the laboratory involved. When a sponsor uses the services of a contract laboratory, consulting laboratory, contractor, or grantee to conduct all or any portion of a nonclinical laboratory study, it is advisable to obtain the written consent of such groups to submit to inspection by the FDA on request as a condition of placing work with the contractor or grantee. This is especially true in the case of contractors or grantees who do not routinely conduct nonclinical laboratory studies and may be unaware of their obligation to permit such inspection; they may not be inclined to consent to inspection.

If a testing facility refuses to permit an FDA inspection, none of the nonclinical laboratory studies or parts of studies conducted by that laboratory will be considered in support of an application for a research or marketing permit. The results of such studies must be submitted to FDA, but the results would not be accepted as evidence of the safety of the test article. Such results could be used by FDA to support a finding that the test article was not safe, however.

Inspections by the FDA must occur at "reasonable times," which is generally defined as during normal business hours. Inspections must also be conducted in a "reasonable manner," which would include adherence by the inspector to all laboratory safety policies (e.g., wearing safety goggles) and compliance with normal requirements for donning protective apparel (gown or lab coat, hat, mask, shoe covers, etc.) before entering animal housing areas.

The inspection authority of the FDA includes the right to copy records and to collect samples. It is discretionary with the inspected laboratory whether to charge for copies of records or to provide them to the inspector free of charge. The same is true with regard to the inspector's request for samples, although requests for samples are rare during GLP inspections. In most cases, laboratories will provide samples and copies of documents free of charge unless FDA requests for these are excessive.

Quality assurance unit records are exempt from routine FDA inspection and copying authority on the theory that such records are more likely to be complete and candid if they are exempt from review by the FDA. This exemption extends only to records of QA inspection and audit findings and records of corrective actions recommended and taken. All other QA records are subject to inspection and copying by FDA.

The one exception to FDA's policy of not seeking access to QA records of findings and problems or of corrective actions recommended and taken is that FDA may seek production of these reports in litigation under applicable procedural rules. The QAU should therefore seek the advice of house counsel as to the retention period for such records.

Before 1992 FDA normally provided at least 1 weekly advance notice of a GLP inspection except in the case of for-cause inspections, which usually occurred without advance notice. Current FDA policy, however, is to conduct all GLP inspections in the United States without advance notice. Generally FDA continues to provide advance notice of GLP inspections outside the United States.

For an excellent discussion of the legal issues surrounding FDA's inspectional authority, see volumes 16 and 52 of this series [13, 14].

SUBPART B: ORGANIZATION AND PERSONNEL

§ 58.29: Personnel

(a) Each individual engaged in the conduct of or responsible for the supervision of a nonclinical laboratory study shall have education, training, and experience, or combination thereof, to enable that individual to perform the assigned functions.
(b) Each testing facility shall maintain a current summary of training and experience and job description for each individual engaged in or supervising the conduct of a nonclinical laboratory study.

The FDA has refrained from specifying exactly what scientific disciplines, education, training, or expertise qualify individuals to participate in the conduct of a nonclinical laboratory study. These factors vary from study to study, and FDA has merely indicated that the question of employee qualifications should be carefully considered by laboratory management. Laboratory management therefore has considerable latitude to define job qualifications. Any reputable laboratory will find it to be in its own best interest to hire competent individuals and to provide adequate on-the-job training to qualify those individuals to perform their assigned duties. The FDA is not likely to make an issue of employee qualifications unless an inspection reveals an obvious case of employee incompetence.

Documentation of employee qualifications should include at a minimum an educational history for each employee, an employee's employment history to the extent that prior employment has a bearing on the employee's competence to perform his or her current job assignment, and a description of any additional on-the-job training provided to the employee. Any format is acceptable for documentation of employee qualifications as long as all relevant information is included. The degree of detail associated with documentation of on-the-job training varies widely from laboratory to laboratory. Some laboratories document supervisor/trainer sign-off for completion of training in each element of an employee's current job description. Other laboratories merely document successful completion of an employee's initial probationary period. Documentation should be updated periodically to reflect changes in educational background and any additional training provided to the employee.

> (c) There shall be a sufficient number of personnel for the timely and proper conduct of the study according to the protocol.

The requirement for adequate numbers of personnel was included in the GLP regulations as a result of FDA's pre-GLP inspection of a laboratory that had taken on more work than its employees could properly perform. The result, according to FDA, was poor-quality or even fraudulent data.

In FDA's opinion, a shortage of qualified personnel can lead to inadequate or incomplete monitoring of a study, delayed preparation and analysis of the study, and delayed preparation and analysis of the study results. The numbers of personnel conducting a study should be sufficient to avoid such problems.

Today it is unlikely that a laboratory would be prospectively cited by FDA for inadequate numbers of personnel. Any citation in this area is more likely to be retrospective and based on actual evidence of poor quality work related to inadequate numbers of personnel.

> (d) Personnel shall take necessary personal sanitation and health precautions designed to avoid contamination of test and control articles and test systems.

(e) Personnel engaged in a nonclinical laboratory study shall wear clothing appropriate for the duties they perform. Such clothing shall be changed as often as necessary to prevent microbiological, radiological, or chemical contamination of test systems and test and control articles.

Although these sections of the GLPs are designed to protect test and control articles and test systems, laboratory management should also take into account federal and state requirements for the protection of the health and safety of the employees. The minimum acceptable protective apparel for employees working with test and control articles and with animals is a laboratory coat over street clothing. Many laboratories provide uniforms. A sufficient supply of clean apparel should be provided by the company to allow frequent changes if suggested by the hazards of the materials or if necessary to protect against cross-contamination. The wearing of hats, gloves, masks, and shoe covers (preferably of the disposable variety) is highly recommended. Enough of these items should be provided to permit changes when moving between rooms. Safety glasses or protective goggles will be appropriate for some hazardous operations.

A laboratory should have a generic policy for the safe handling of chemicals plus special policies for work with hazardous materials.

A laboratory should have a generic policy for the safe handling of chemicals plus special policies for work with hazardous materials.

Refer to NIH publication no. 85–23, *Guide for the Care and Use of Laboratory Animals*; NIH publication no. 81–2385, *NIH Guidelines for the Laboratory Use of Chemical Carcinogens*; and the Public Health Service's *Biosafety Guidelines for Microbiological and Biomedical Laboratories* for additional discussion of these issues.

(f) Any individual found at any time to have an illness that may adversely affect the quality and integrity of the nonclinical laboratory study shall be excluded from direct contact with test systems, test and control articles and any

> other operation or function that may adversely affect the study until the condition is corrected. All personnel shall be instructed to report to their immediate supervisors any health or medical conditions that may reasonably be considered to have an adverse effect on a nonclinical laboratory study.

The potential for spreading disease organisms from animals to humans and vice versa is not obvious to most people. These so-called zoonotic diseases include agents of all the major categories of infectious organisms: viruses, bacteria, parasites, and fungi. Infectious hazards are insidious, and therefore safe practices should be habitual and strictly enforced. All employees should be instructed as to the nature of these hazards and the means to take to protect animals and themselves from infection. Employees should also be instructed to report all personal illnesses to their supervisor. The supervisor can then determine whether or not it would be appropriate for the employee to have contact with test and control articles and test systems.

§ 58.31: Testing Facility Management

For each nonclinical laboratory study, testing facility management shall

> (a) Designate a study director as described in § 58.33, before the study is initiated.
> (b) Replace the study director promptly if it becomes necessary to do so during the conduct of a study.

A study director can be designated for each study in the study protocol that is approved by management or in separate documentation that is signed by management. As mentioned in the discussion of GLP definitions, only one person may be designated as study director. Study codirectors are not permissible, but an alternate study director may be designated. If the study director must be replaced, this may be accomplished by protocol amendment (if the original study director was designated in the protocol) or by separate documentation (if separate documentation was used to appoint the original study director).

(c) Assure that there is a quality assurance unit as described in § 58.33, before the study is initiated.
(d) Assure that test and control articles or mixtures have been appropriately tested for identity, strength, purity, stability, and uniformity, as applicable.
(e) Assure that personnel, resources, facilities, equipment, materials, and methodologies are available as scheduled.
(f) Assure that personnel clearly understand the functions they are to perform.
(g) Assure that any deviations from these regulations reported by the quality assurance unit are communicated to the study director and corrective actions are taken and documented.

These duties, which are more administrative than scientific, are the responsibility of management. "Management" will generally be defined as the person or persons who have authority within an organization to effect whatever changes are necessary to assure that these duties are adequately discharged. Identification of such persons will vary, depending on the structure of each organization. Management may, of course, delegate these duties to others within the organization. Responsibility, however, continues to reside with the person(s) with the authority to effect change.

The requirement to assure that deviations reported by the QAU are communicated to the study director and that corrective actions are taken and documented does not mean that management itself must communicate the findings and take appropriate corrective action. An efficient QAU will document deviations and the fact that corrective action has already occurred in reports that are distributed to both management and the study director. The need for additional management follow-up will then be necessary only in those few instances in which corrective action was not adequately negotiated between the QAU and the scientific staff before the issuance of the QAU report. When corrective action is underway but not complete at the time of the QAU report, the report need only indicate that fact with additional follow-up provided in subsequent reports.

§ 58.33: Study Director

For each nonclinical laboratory study, a scientist or other professional of appropriate education, training, and experience, or combination thereof, shall be identified as the study director. The study director has overall responsibility for the technical conduct of the study, as well as for the interpretation, analysis, documentation and reporting of results, and represents the single point of study control. The study director shall assure that

(a) The protocol, including any change, is approved as provided by § 58.120 and is followed.

The study director does not approve the protocol but only makes certain that approval is obtained from sponsor management.

(b) All experimental data, including observations of unanticipated responses of the test system, are accurately recorded and verified.

The study director is not required to observe every data collection event, but should assure that data are collected as specified by the protocol and the standard operating procedures (SOPs) and that data collection includes the accurate recording of unanticipated responses of the test system. The study director should also review data periodically, or assure that such review occurs, to promote the accurate recording of data and to assure that data are technically correct.

(c) Unforeseen circumstances that may affect the quality and integrity of the nonclinical laboratory study are noted when they occur, and corrective action is taken and documented.

Systems must be in place to assure that the study director is promptly notified of unforeseen circumstances that may have an effect on the integrity of the study. The study director must then assure that corrective action is taken and documented in response to those unforeseen circumstances.

(d) Test systems are as specified in the protocol.

The determination of the appropriateness of the test system is a scientific decision made by management at the time of protocol approval. The study director need only assure that protocol specifications are followed.

(e) All applicable good laboratory practice regulations are followed.

This section suggests the need for frequent interaction between the study director and QA personnel. Deviations from GLP requirements noted by the QAU must be reported periodically to management and the study director. If those reports indicate that corrective action is still needed for any deviation from regulatory requirements, it is the study director's responsibility to assure that corrective action occurs.

The study director's role is not simply to react to reports of regulatory deviations from the QAU, but also to play a proactive role to assure that study personnel are aware of GLP requirements and that deviations from those requirements do not occur.

(f) All raw data, documentation, protocols, specimens, and final reports are transferred to the archives during or at the close of the study.

Materials may be transferred to the archives as the study progresses or at the close of the study. Although FDA has defined the close of the study as the time in which the final report of the study is signed by the study director, it will not be a violation of regulatory requirements if materials reach the archives in a reasonable period of time after the signature date.

§ 58.35: Quality Assurance Unit

(a) A testing facility shall have a quality assurance unit which shall be responsible for monitoring each study to

> assure management that the facilities, equipment, personnel, methods, practices, records, and controls are in conformance with the regulations in this part. For any given study the quality assurance unit shall be entirely separate from and independent of the personnel engaged in the direction and conduct of that study.

Arguments for maintaining a full-time staff of QA professionals have been previously delineated in the discussion of the definition of the QAU.

> (b) The quality assurance unit shall:
>
> (1) Maintain a copy of a master schedule sheet of all nonclinical laboratory studies conducted at the testing facility indexed by test article and containing the test system, nature of study, date study was initiated, current status of each study, identity of the sponsor, and name of the study director.

The FDA believes that maintenance of a detailed master schedule sheet is essential to the proper functioning of the QAU. In actual practice few QA groups use the master schedule in the performance of QA functions. Few do more than maintain a master schedule for the benefit of FDA inspectors, who use it to gauge the volume of GLP-regulated work being conducted by a laboratory and to aid in the random selection of studies for review during an inspection.

There is no requirement for the QAU to prepare the master schedule. The master schedule may be prepared by some organizational unit other than the QAU as long as the QAU maintains a copy in its files.

The FDA has indicated that a study should first appear on the master schedule on the date the protocol is signed by the study director. A study may come off the master schedule when the final report of the study is signed by the study director.

In the preamble (¶ ¶ 8 and 13) to the 1987 GLP revisions [4] the master schedule was referred to as "raw data." In a subsequent clarification, Dr. Paul Lepore indicated that the term raw data had appeared in quotes in the preamble to indicate that the term was not being used as defined in § 58.3

(k) of the regulations; rather, the term was used to emphasize that copies of the master schedule were subject to the record retention requirements of §§ 58.190 and 58.195.

Additional language in the preamble (¶ 15) to the GLP revisions of 1987 [4] as well as enforcement policies of individual FDA investigators have broadly interpreted the requirement to include the "current status of each study" on the master schedule. According to this view, the master schedule should include such study events as test article-mixture preparation, test system dosing, and in-life observation. Because such detailed information is usually available in other study documentation (e.g., protocol, study schedules), most laboratories limit a description of current status to broad categories such as in-life phase, "study terminated," "report preparation," and "report issuance."

It is permissible to identify the sponsor on the master schedule by code rather than by name. This allows a contract laboratory to protect client confidentiality if the master schedule is examined by one of many clients. The contract laboratory must, however, make the names of sponsors available to FDA upon request.

Many laboratories maintain the master schedule on computer, and find it a helpful tool for the allocation of resources and the scheduling of work. A computerized master schedule can also provide the index of archive materials required by § 58.190(e) of the regulations.

(2) Maintain copies of all protocols pertaining to all nonclinical laboratory studies for which the unit is responsible.

A proper discharge of QAU responsibilities requires a knowledge of protocol requirements. One of the QAU's responsibilities is to inspect study conduct to assure that there are no deviations from protocol requirements. Preparation for and conduct of those inspections require ready access to a copy of the protocol and all protocol amendments.

(3) Inspect each nonclinical laboratory study at intervals adequate to assure the integrity of the study and maintain

> written and properly signed records of each periodic inspection showing the date of the inspection, the study inspected, the phase or segment of the study inspected, the person performing the inspection, findings and problems, action recommended and taken to resolve existing problems, and any scheduled date for reinspection. Any problems found during the course of an inspection which are likely to affect study integrity shall be brought to the attention of the study director and management immediately.

"Inspect" has been defined by FDA to mean an actual examination and direct observation of the facilities and operations for a given study while the study is in progress and not merely a review of the records of a study. The QAU function is to observe and report on the state of compliance of a study with the requirements of the study protocol, laboratory SOPs, and the GLP regulations. The QAU role is not just to verify the results of a study.

Each QAU may exercise reasonable flexibility and judgment to establish an inspection schedule that it believes is "adequate to assure the integrity of the study." The FDA has indicated, however, that every study must be inspected in process at least once. Additional inspections may be randomly scheduled in such a way that over a series of studies each phase for each type of study is inspected. Any random sampling approach to inspections should be statistically based and should be described and justified in the QAU's SOPs.

Under U.S. Department of Agriculture animal welfare regulations [15], nonclinical laboratory studies in animals must be reviewed and approved by the testing facility's Institutional Animal Care and Use Committee (IACUC). The FDA has indicated that IACUC review is part of the conduct of a nonclinical laboratory study and therefore IACUC activities should be periodically inspected by QAU. Because IACUC activities are subject to QAU inspection, FDA has indicated that a member of the QAU may not serve as a voting member of the IACUC but may serve as a nonvoting member.

The information to be recorded in QAU inspection records is straightforward, as is the requirement for the QAU to immediately report significant problems to management and the study director.

(4) Periodically submit to management and the study director written status reports on each study, noting any problems and the corrective actions taken.

The frequency of the QAU's periodic reports to management is left to the discretion of the laboratory. Reports at intervals of approximately amonth are fairly standard within the regulated community. The description of problems noted during QAU inspections need not be extremely detailed unless the problems remain uncorrected. The primary purpose of the report is to assure management that study quality is being maintained and that management intervention is not required.

(5) Determine that no deviations from approved protocols or standard operating procedures were made without proper authorization and documentation.

As noted previously, QAU review of adherence to protocol requirements and SOP are part and parcel of the inspection process.

(6) Review the final study report to assure that such report accurately describes the methods and standard operating procedures, and that the reported results accurately reflect the raw data of the nonclinical laboratory study.

The QAU audit should verify the accuracy and completeness of data and information presented in the final report. The audit should include the narrative description of materials, methods, and results as well as all tabulated data.

For critical study data (e.g., microscopic pathology data and data on tumor incidence) a QAU may elect to perform a 100% audit. For other data a random sampling approach to the audit is perfectly acceptable. Any such random sampling program should be statistically based [16, 17].

For reasons described in the discussion of § 58.185(c), the QAU will normally audit the final draft of the report before it is signed by the study director.

> (7) Prepare and sign a statement to be included with the final study report which shall specify the dates inspections were made and findings reported to management and to the study director.

The list of inspection dates in the QAU statement may not be sufficient to reveal the extent of the QAU audit and inspectional activity for any given study (e.g., when several inspections of a study occur on the same date). For this reason, some laboratories also list the study phases that were inspected even though this is not required by the regulations.

When a random sampling approach to the inspection process is used, it may be desirable to indicate the date(s) of inspection(s) of similar studies during a period that includes the time of conduct of the study for which the QAU statement is being prepared. Any such additional inspections should be clearly labeled as such.

> (c) The responsibilities and procedures applicable to the quality assurance unit, the records maintained by the quality assurance unit, the records maintained by the quality assurance unit, and the method of indexing such records shall be in writing and shall be maintained. These items including inspection dates, the study inspected, the phase or segment of the study inspected, and the name of the individual performing the inspection shall be made available for inspection to authorized employees of the Food and Drug Administration.

The QA SOP manual should describe QAU audit and inspection techniques with attached inspection checklists, if used. Statistically based methods for random selection of phases of studies for inspection and for random selection of data points during final report audits should be described and justified. Any designation of study phases as "critical" or "noncritical" used to establish the frequency of study inspections should also be described and justified. The SOP manual should also describe the method for communicating audit and inspection findings to the study director and management, including a definition of who receives an copy of the reports. Finally, the SOP manual should describe QAU record-filing systems and the method for indexing those records. For filing and indexing systems, the QAU will find it most efficient to base its filing system on the study numbering system used by the safety testing laboratory. It can then utilize the safety laboratory's archive index system for indexing QAU records. The indexing system for QAU records should permit speedy access to such records in the event of any FDA request to review those records during an FDA inspection. The FDA may review and copy any QAU records except those excluded by § 58.15(a).

> (d) A designated representative of the Food and Drug Administration shall have access to the written procedures established for the inspection and may request testing facility management to certify that inspections are being implemented, performed, documented, and followed-up in accordance with this paragraph.
> (Collection of information requirements approved by the Office of Management and Budget under number 0919–0203)

As previously mentioned, QAU records of findings and problems and of corrective actions recommended and taken are exempt from routine FDA inspection. To compensate for this lack of routine inspectional authority, the FDA has access to the QAU's written procedures. The FDA may review QAU written procedures to judge the adequacy of inspection schedules and to determine whether or not systems are in place for communicating inspection findings to management personnel. The FDA may also request facility management to certify in writing that inspections are being implemented, performed, documented, and followed up in accordance with GLP requirements.

SUBPART C: FACILITIES

§ 58.41: General

> Each testing facility shall be of suitable size and construction to facilitate the proper conduct of nonclinical laboratory studies. It shall be designed so that there is a degree of separation that will prevent any function or activity from having an adverse effect on the study.

If a testing facility is too small to handle the volume of work it has set out to do, there may be an inclination to mix incompatible functions. Examples might include the simultaneous conduct of studies with incompatible species (e.g., old world primates and new world primates) in the same room, setting up a small office in the corner of an animal housing area, housing an excessive number of animals in a room, or storing article/carrier mixtures in an animal room.

The facility should be constructed of materials that facilitate cleaning. Heating, ventilation, and air conditioning (HVAC) systems should be of adequate capacity to produce environmental conditions that comply with employee and animal health and safety standards and should be designed to prevent cross-contamination.

The location of a facility (e.g., next to a farm in which pesticides, herbicides, and fertilizers are frequently used or next door to a chemical factory that generates noxious fumes) could have an adverse effect on the conduct of a nonclinical laboratory study unless the facility is designed to protect against outside environmental contaminants. Although the GLP revisions of 1987 eliminated "location" as a consideration in § 58.41, it is still a strong consideration in the design and construction of nonclinical laboratories.

Facilities should be designed to avoid disturbances such as intermittent or continuous noise from within or outside the facility, frequent traffic in and out of animal rooms, obnoxious odors (e.g., chemical odors that are carried by ventilation systems from laboratories to animal housing areas), and animal disturbances, which can be caused by facility design factors. Extreme care should be taken to design special protection for those animals (e.g., pregnant animals) that are especially sensitive to interfering disturbances.

In short, FDA is concerned that a facility be designed and constructed to ensure the adequacy of the facility for conducting nonclinical laboratory studies and to ensure the quality and integrity of study data.

§ 58.43: Animal Care Facilities

> (a) A testing facility shall have a sufficient number of animal rooms or areas, as needed, to assure proper: (1) Separation of species or test systems, (2) isolation of individual projects, (3) quarantine of animals, and (4) routine or specialized housing of animals.

Note the words as needed and proper. The facility's veterinarian in charge should be consulted as to when generally accepted standards for laboratory animal care require the separation, isolation, or specialized housing of animals. It is generally accepted that all newly received animals should undergo a quarantine and acclimation period.

"Isolation" generally connotes a setting apart, by use of physical barriers, from all other projects. "Separation," on the other hand, can be accomplished by spatial arrangements (e.g., two projects can be assigned to different parts of the same room).

> (b) A testing facility shall have a number of animal rooms or areas separate from those described in paragraph (a) of this section to ensure isolation of studies being done with test systems or test and control articles known to be biohazardous, including volatile substances, aerosols, radioactive materials, and infectious agents.

A laboratory involved in work with the hazardous materials described in this section also needs to be familiar with regulations of the Occupational Safety and Health Administration (or state equivalent), the Department of Agriculture, and the Nuclear Regulatory Commission, all of which have a role in the regulation of such materials.

> (c) Separate areas shall be provided, as appropriate, for the diagnosis, treatment, and control of laboratory animal diseases. Even if the laboratory does not have such a policy, there may be instances (e.g., non-contagious diseases) where diseased animals need not be isolated for treatment. Whether or not to treat and whether or not to isolate is a scientific decision which should be made by the study director in consultation with other scientific personnel.

If a laboratory's policy is to euthanize all diseased animals, it need not provide separate areas for the diagnosis, treatment, and control of laboratory animal diseases. Even if the laboratory does not have such a policy, there may be instances (e.g., noncontagious diseases) in which diseased animals need not be isolated for treatment. Whether or not to treat and whether or not to isolate is a scientific decision that should be made by the study director in consultation with other scientific personnel.

If a laboratory intends to treat rather than euthanize diseased animals, it is best to have an area separate from other animal housing and holding areas for the isolation of diseased animals (if this is deemed necessary). A second area may be needed to treat animals with contagious diseases separately from those animals being treated for noncontagious diseases.

> (d) When animals are housed, facilities shall exist for the collection and disposal of all animal waste and refuse or for safe sanitary storage of waste before removal from the testing facility. Disposal facilities shall be so provided and operated as to minimize vermin infestation, odors, disease hazards, and environmental contamination.

A laboratory may dispose of animal waste and refuse onsite (e.g., incineration) or may use a contract service for pickup and disposal. Some animal waste and refuse may meet EPA's definition of hazardous waste (e.g., waste or refuse from animals treated with hazardous materials or animals carrying infectious diseases) and must be disposed of in compliance with EPA regulations issued under the RCRA. Waste and refuse from animals treated with radioactive materials must be disposed of in compliance with regulations of the Nuclear Regulatory Commission.

Containers with tight-fitting lids should be used for the temporary storage of animal waste and refuse before disposal to minimize vermin infestation, odors, disease hazards, and environmental contamination.

§ 58.45: Animal Supply Facilities

> There shall be storage areas, as needed, for feed, bedding, supplies, and equipment.
>
> Storage areas for feed and bedding shall be separated from areas housing the test systems and shall be protected against infestation or contamination. Perishable supplies shall be preserved by appropriate means.

Animal feed and bedding should never be stored in areas in which animals are housed. It is also contrary to good animal husbandry practices to store supplies and equipment in animal housing areas.

Animal feed and bedding should be stored off the floor to facilitate cleaning. Food storage areas and areas used to store other perishable supplies should be temperature-controlled to protect against deterioration of the stored materials.

The first line of defense against vermin should be perimeter control, that is, controls to prevent the entry of vermin into a facility. If vermin control within the facility is necessary, care should be taken to protect supplies of feed and bedding from contamination by vermin control materials.

§ 58.47: Facilities for Handling Test and Control Articles

> (a) As necessary to prevent contamination or mix-ups, there shall be separate areas for:

(1) Receipt and storage of the test and control articles.
(2) Mixing of the test and control articles with a carrier, e.g., feed.
(3) Storage of the test and control article mixtures.

(b) Storage areas for the test and/or control article and test and control mixtures shall be separate from areas housing the test systems and shall be adequate to preserve the identity, strength, purity, and stability of the articles and mixtures.

The twin goals of § 58.47 are to prevent cross-contamination and mix-ups. Facility management must provide the necessary degree of separation to meet these goals. Separate rooms are not required for each of the described functions if adequate separation can be provided by spatial arrangements within a room, by special air-handling techniques, and/or by strictly enforced procedural requirements.

Dedicated areas are usually provided for the receipt and storage of test and control articles. Such articles are usually stored under lock and key. Areas for weighing test and control articles are often equipped with special air-handling systems, sometimes roomwide and other times limited to the area immediately surrounding the weighing devices. Many laboratories have a policy for weighing only one test or control article at any one time in any one area.

Operations with high cross-contamination potential (e.g., mixtures of test or control articles with animal diets) are often conducted in small, dedicated, individual cubicles equipped with special and separate air-handling systems or are conducted under a fume hood. Special mixing equipment (e.g., enclosed twin-shell blenders) can be used to reduce the chance of cross-contamination.

If it is necessary to store test and control article mixtures, such materials should be stored entirely separately from animal housing areas. Special storage conditions (e.g., refrigeration and protection from light) must be available if needed to preserve and maintain the quality and stability of the mixtures.

§ 58.49: Laboratory Operation Areas

Separate laboratory space shall be provided, as needed, for the performance of the routine and specialized procedures required by nonclinical laboratory studies.

A laboratory must provide adequate and, if necessary, separate space for the performance of routine and specialized procedures. Examples of specialized procedures include aseptic surgery, necropsy, histology, radiography, handling of biohazardous materials, and cleaning and sterilizing of equipment and supplies.

§ 58.51: Specimen and Data Storage Facilities

Space shall be provided for archives, limited to access by authorized personnel only, for the storage and retrieval of all raw data and specimens from completed studies.

A laboratory that conducts nonclinical laboratory studies must provide space for the storage of raw data and specimens from such studies. Access to the archives must be controlled. This is best accomplished by providing a lockable area and by defining in the laboratory's SOPs who has access to archive materials and under what conditions (e.g., use only within the archives or "check-out" rights).

Raw data and specimens need not be transferred to the archives until the completion of the study. Many laboratories elect to transfer material to the archives as it is completed, however, to provide greater data security. The FDA has stated that all materials must be transferred to the archives within a reasonable period of time after the study director signs the final report.

See the discussion of § 58.190 for other archive requirements.

SUBPART D: EQUIPMENT

§ 58.61: Equipment Design

Equipment used in the generation, measurement, or assessment of data and equipment used for facility environmental

> control shall be of appropriate design and adequate capacity to function according to the protocol and shall be suitably located for operation, inspection, cleaning, and maintenance.

Equipment used to generate, measure, or assess data should undergo a validation process to ensure that such equipment is of appropriate design and adequate capacity and will consistently function as intended. Examples of such equipment include scales; balances; analytical equipment (HPLC, GC, etc.); hematology, blood chemistry, and urine analyzers; computerized equipment for the direct capture of data; and computers for the statistical analysis of data. Because the data generated, measured, or assessed by such equipment are the essence of a nonclinical laboratory study, the proper functioning of such equipment is essential to valid study results.

Safety assessment scientists and technicians and even QA personnel sometimes overlook the importance of environmental control equipment to valid study results. Animals stressed by extremes of temperature or humidity may yield spurious data; reproductive toxicology studies may be compromised by malfunctioning timers for the control of light/dark cycles; inadequate air filtration may expose experimental animals to environmental contaminants that confound experimental results.

All equipment described in § 58.61 should be located in such a manner as to promote proper operation, inspection, cleaning, and maintenance.

§ 58.63: Maintenance and Calibration of Equipment

> (a) Equipment shall be adequately inspected, cleaned, and maintained. Equipment used for the generation, measurement, or assessment of data shall be adequately tested, calibrated and/or standardized.

The need for regular inspection, cleaning, and maintenance of equipment is well recognized in the scientific community. A laboratory should establish schedules for such operations based on the manufacturer's recommendations and laboratory experience. In most instances these schedules will be defined as to periodicity, although in some cases an "as needed" schedule will be acceptable.

The terms *test*, *calibration*, and *standardization* are interrelated. Each term has a special meaning, but there is some overlap of the terms.

Test can be defined as an examination of an item or system to determine compliance with its specifications. Under this definition test would include operations to calibrate or standardize but would also include the process of total system validation.

Calibration has been defined as a comparison of a measurement standard or instrument of known accuracy with another standard or instrument to detect, correlate, report, or eliminate by adjustment any variation in the accuracy of the item being compared [18].

Standardization is a comparison with a standard of known and accepted value. Standards may be of several sources: primary standards [prototype state-of-the-art standards found at NIST; the National Institute of Standards and Technology (formerly known as the National Bureau of Standards; (NBS), or national equivalent outside the United States]; secondary, working standards (standards calibrated to primary standards, which are used for working tools and instruments); and in-house-developed or interim standards (standards developed and used by a particular facility when no primary standard is available).

Scales and balances should be calibrated at regular intervals, usually ranging from 1 month to 12 months, depending on manufacturers' recommendations, laboratory experience, and the extent of use. Intervals should be selected with a recognition that if a scale or balance is found to be out of calibration, it will cast doubt on the accuracy of every weight measured by that scale or balance since the last calibration. Scales and balances should also be standardized with a range of standard weights at frequent intervals. Many laboratories standardize scales and balances before each use, and some also standardize at periodic intervals during each use. The range of standard weights should bracket the expected experimental values. Standard weights should be traceable to NIST standards and should themselves be periodically calibrated.

The use of standard solutions, reference standards, and quality control samples, whether prepared by the laboratory or purchased commercially, is essential to valid analyses of test and control article/carrier mixtures and biological fluids (blood, serum, plasma, etc).

A Ph meter should be standardized before each use according to directions in the manufacturer's manual.

Electrocardiographs usually have a built-in facility for generating an electrical impulse of known intensity. This facility should be used during the recording of electrocardiograms to check periodically on the proper functioning of the equipment.

Heating, ventilation, and air conditioning equipment should be regularly inspected and maintained. Filters on environmental control equipment should be inspected on a regular basis and changed as needed.

> (b) The written standard operating procedures required under § 58.81(b)(11) shall set forth in sufficient detail the methods, materials, and schedules to be used in the routine inspection, cleaning, maintenance, testing, calibration and/or standardization of equipment, and shall specify, when appropriate, remedial action to be taken in the event of failure or malfunction of equipment. The written standard operating procedures shall designate the person responsible for the performance of each operation.

All aspects of a laboratory's program for the routine inspection, cleaning, maintenance, testing, calibration, and/or standardization of equipment must be in writing (i.e., SOPs, supplemented as necessary by equipment manuals). This would include a description of cleaning materials; inspection, cleaning, and maintenance methods and schedules; calibration and standardization methods and parameters; and the job title of personnel responsible for the performance of each operation. Specification of remedial actions to be taken in response to equipment failure or malfunction should be as comprehensive as possible. Common troubleshooting problems with appropriate remedial action are frequently included in equipment manufacturers' manuals, which can be cited in the SOPs. For other types of problems, it will generally be sufficient to indicate in the SOPs that professional assistance will be enlisted (e.g., manufacturers' repair services).

Copies of equipment SOPs must be easily and readily accessible by laboratory personnel.

"When appropriate" means that a laboratory only need specify remedial action in response to equipment failure or malfunction when remedial action is appropriate to the piece of equipment. A laboratory may elect to discard rather than repair faulty equipment; however, records for the discarded equipment, including records of previous maintenance and calibration, must be retained for the length of time described in § 58.195(b) and (f).

> (c) Written records shall be maintained of all inspection, maintenance, testing, calibrating and/or standardizing operations. These records, containing the date of the operation, shall describe whether the maintenance operations were routine and followed the written standard operating procedures. Written records shall be kept of non-routine repairs performed on equipment as a result of failure and malfunction. Such records shall document the nature of the defect, how and when the defect was discovered, and any remedial action taken in response to the defect.
> (Collection of information requirements approved by the Office of Management and Budget under number 0919–0203)

As with any activity required by regulation, records must be maintained of all equipment inspection, maintenance, testing, calibrating, and/or standardizing operations. The records required by this section of the regulations are necessary to the reconstruction of a study and provide the FDA with added assurance as to the validity and integrity of data. The FDA has indicated, however, that it is not necessary to maintain records of cleaning operations on the theory that the costs of maintaining such records exceeds the benefits.

Records of routine maintenance operations may reference the SOPs for a description of the operations. For nonroutine repairs in response to equipment failure or malfunction, repair records must contain the following detailed information: nature of the defect, how the defect was discovered, when the defect was discovered, and remedial action taken in response to the defect. Remedial action should include a review of possible effects on data generated before the defect was discovered. Because repairs are likely to involve repairmen from outside the laboratory, care must be taken to ensure that such persons provide full documentation of the nature of the problem and remedial action taken in response to the problem.

Equipment inspection, maintenance, and repair records can be recorded in a logbook especially designed for that purpose. For equipment that is moved from laboratory to laboratory, the logbook should accompany the equipment when it is moved. Documentation of calibrating or standardizing operations, on the other hand, may be more efficiently recorded with the associated records of the data acquisition activities.

SUBPART E: TESTING FACILITIES OPERATION

§ 58.81: Standard Operating Procedures

(a) A testing facility shall have standard operating procedures in writing setting forth nonclinical laboratory study methods that management is satisfied are adequate to insure the quality and integrity of the data generated in the course of a study. All deviations in a study from standard operating procedures shall be authorized by the study director and shall be documented in the raw data. Significant changes in established standard operating procedures shall be properly authorized in writing by management.

Preparation of written SOPs was a major undertaking for most GLP-regulated laboratories. Keeping SOP manuals up to date continues to be a major effort for these labs. To assure that SOP manuals remain up to date, many laboratories have a policy for mandatory, periodic review (and update, if necessary) of all SOPs.

Study protocols define "what" is to be done during the course of a study; SOPs define "how" to carry out protocol-specified activities.

There are many acceptable formats for SOPs. The author prefers an activity-oriented format, written in playscript style and including a designation of the actor (i.e., who is responsible for the activity) and a chronological listing of action steps (i.e., of what the activity consists) [19]. (See Fig. 1 for a sample of an SOP in this format.) Prime consideration should be given to making the SOP manual user-friendly so that it is a document which invites rather than discourages routine usage by those responsible for performing tasks in compliance with the SOP.

In the preparation and revision of SOPs, a major consideration is the degree of detail to be incorporated into SOPs. As a general rule, SOPs should be detailed enough to provide meaningful direction to study personnel for the conduct of routine laboratory activities. In determining the level of detail, it is acceptable to take into consideration the education, training, and experience of the personnel who will be responsible for those activities. For example, an analytical procedure to be carried out by a trained chemist would instruct the chemist to pipette 5 ml of a reagent, but need not provide detail of how to pipette. It is generally not advisable to specify suppliers of materials in SOPs because suppliers may change frequently. It is always advisable to allow for a range of acceptable approaches to any procedure if a more specific, restrictive, and defined activity is not necessary to assure study quality. If written too restrictively, SOPs are fre-

PARKE-DAVIS	PROCEDURES	VOLUME:
PHARMACEUTICAL RESEARCH DIVISION		SECTION: 80.745.05 80.745
DEPARTMENT OF PATHOLOGY & EXPERIMENTAL TOXICOLOGY		EXHIBIT:
SUBJECT: TEST & CONTROL ARTICLES - STABILITY RE-ANALYSIS		

PURPOSE

The following provisions elaborate upon the test and control article policy (80.745.02), establishing specific responsibilities and delegations of authority involved in the periodic re-analysis of test and control articles for assurance of stability.

LOCATION AFFECTED

Pathology and Experimental Toxicology, Ann Aarbor, Michigan.

PROCEDURE

RESPONSIBILITY	ACTION	
Study Director or Scientist/Technician Designee	1.	Review the available stability data of test and control articles prior to initiation of a study. Arrange for periodic re-analysis of each lot if the stability of the test or control article cannot be determined before initiation of a study or available
	2.	At termination of studies of less than 6 months' duration (including genetic toxicology and reproductive toxicology investigations), select a sample (approximately 200 mg if possible) of test or control article for potency re-analysis.
NOTE:	Samples for re-analysis should be taken from the container in use at the time.	
	3.	If the study is to last more than 6 months, forward sample for re-analysis at least every 6 months.
NOTE:	If studies with the same test and control articles are conducted concurrently, ONE POTENCY RE-ANALYSIS SAMPLE COLLECTED AFTER COMPLETION OF THE LAST STUDY WILL SUFFICE. Please refer to Drug Inventory Forms for timing of sample collection.	
	4.	If at any time, regardless of study duration, a lot is nearing depletion, or a lot is being returned and a potency assay has not been conducted, submit a sample for re-analysis.
	5.	Receive the written report, evaluate stability test results and incorporate document into the appropriate study file(s).
	6.	If a significant loss of activity or increase in contaminant(s) has occurred, submit a sample for re-test and notify Departmental Supervision and Quality Assurance immediately (significance of potency loss will vary according to the precision of the assay procedure, but a loss of □10% is cause for concern; in general, no single contaminant should be present at >0.5% and total contaminants should be □2%).
Directors or Section Heads, pathology and Experimental Toxicology and Quality Assurance Representative	7.	If re-test confirms the loss of potency or increase in contaminants, determine how this information affects the associated studies.
DATE: [Effective Date Entered Here]	[Management Approval Signature]	PAGE 1 of 1

Figure 1 Sample SOP.

quently in need of revision. On the other hand, if insufficient detail is included in the SOPs, they fail to provide adequate direction to study personnel. With experimentation and experience a laboratory can strike a reasonable balance between too much and not enough detail. It is always a good idea to solicit comments from those who use the SOP manual (the workers at the bench) in striking this balance.

If an exception to a SOP is to be made for an individual study, that exception must be authorized in writing by the study director, and the written authorization must be maintained with the raw data for the study. If a change in procedure represents a new standard way of doing things, then the SOP should be revised, and the revision approved (e.g., by signature) by laboratory management.

> (b) Standard operating procedures shall be established for, but not limited to, the following:
> (1) Animal room preparation.
> (2) Animal care.
> (3) Receipt, identification, storage, handling, mixing, and method of sampling of the test and control articles.
> (4) Test system observations.
> (5) Laboratory tests.
> (6) Handling of animals found moribund or dead during study.
> (7) Necropsy of animals or postmortem examination of animals.
> (8) Collection and identification of specimens.
> (9) Histopathology.
> (10) Data handling, storage, and retrieval.
> (11) Maintenance and calibration of equipment.
> (12) Transfer, proper placement, and identification of animals.

As suggested by "but not limited to," the list of SOP topics in § 58.81 (b) should be considered illustrative, not comprehensive. Many of the topics (e.g., laboratory tests) might involve 100 or more individual SOP titles. The range of topics for which SOPs are required will be governed by the variety of studies routinely conducted in the laboratory. For each procedure required by each type of study, the laboratory should prepare an SOP describing how that procedure should be performed. If a study activity is not yet "standard" or is intended to be a one-time event, it is acceptable to incorporate a detailed description of the "how-to" for that activity in the study protocol or in a laboratory notebook. If such activities become routine, however, an SOP should be prepared.

In the foregoing discussion of § 58.35(b)(3) it was indicated that FDA considers an IACUC review of nonclinical laboratory studies in animals to be part of the conduct of those studies. The FDA has also indicated that IACUC functions should be described in SOPs (even though the U.S. Department of Agriculture regulations [15] that mandate IACUC review do not require that IACUC functions be described in written SOPs!). The FDA has specifically stated [20] that IACUC SOPs should include the following:

A document from a high-ranking laboratory official that states that the laboratory does not condone or support inhumane treatment of animals and that it is the policy of the laboratory to maintain, hold, and use animals in compliance with all applicable regulations, guidelines, and policies

A description of committee members (including the chair), the number of members and their terms of office, and the procedure for replacing committee members

A definition of a quorum for committee activities

A description of how the committee makes decisions

A description of committee documents, including what items go out with a meeting agenda and what items should be describe in meeting minutes

Some laboratories establish a hierarchy of documents and specify that SOPs describe the approved method for study conduct unless an alternate methodology is described in a study protocol. In such a case, the alternate methodology would only be applicable for a study in which the protocol so provides. Because the study director must sign the protocol, such a system provides an easy method for compliance with § 58.81(a), which requires the study director to authorize all deviations from SOPs in a study.

(c) Each laboratory area shall have immediately available laboratory manuals and standard operating procedures relative to the laboratory procedures being performed. Published literature may be used as a supplement to standard operating procedures.

If SOPs are to provide guidance to study personnel on accepted methods for the conduct of routine study procedures, it follows that they must be readily available to the personnel performing those activities. The SOPs should be available in or near the room in which the activities will occur. The requirement for "immediately available" SOPs is not met if an employee must travel some distance in order to consult the SOP manual. In such a case the employee is more likely to guess, and perhaps guess wrongly, about the proper method for study conduct.

The entire SOP manual need not be immediately available as long as those SOPs that describe procedures to be performed are available.

Published literature (and manufacturers' equipment manuals) may supplement SOPs, but will as a general rule not be an acceptable substitute for SOPs.

Well-prepared SOPs will serve as a good training tool for new employees and will provide a handy "crutch" for experienced personnel whose memory of study methods may need some refreshing. Although not a GLP requirement, many laboratories provide each employee with a complete copy of the SOP manual in addition to providing the mandatory "working copies" of individual SOP titles in each work area.

To meet the requirement for "standard" operating procedures, the laboratory is advised to develop a system to ensure that all working copies of the SOPs are identical. When SOP revisions are distributed, all holders of the manual should be instructed at a minimum to destroy the outdated version of the procedure. A better approach is to require the outdated version to be returned to and accounted for at a central location. Follow-up should then be provided to ensure the return of all copies of the outdated procedure. Ideally each distributed copy of the SOPs should be uniquely numbered, and employees should be instructed not to make copies of any individual SOP. This provides better control over the distribution process and helps ensure that all outdated versions of SOPs are destroyed.

With the advance in computer technology and the increased use of computer networks, many laboratories are making SOPs available in electronic form via read-only central computer files. Electronic SOPs help ensure that all personnel are using the current version of an SOP, reduce or eliminate the need for distribution of paper copies of the SOPs, and reduce or eliminate the need for follow-up to ensure that SOP manuals are updated properly. A master, hard copy version of the SOPs that is authorized, signed, and dated by management still must be retained in the archives, and the historical file of SOPs should also contain hard copy versions that have been authorized, signed, and dated by management. Like hard copy SOPs, electronic SOPs must be readily available to study personnel.

> (d) A historical file of standard operating procedures, and all revisions thereof, including the dates of such revisions, shall be maintained.

The historical file of SOPs documents what SOPs were in effect at any time during a laboratory's history. Because FDA inspection of a study often occurs years after the completion of that study, the historical file of SOPs will be of special use to an FDA inspector. Including the effective date on the SOP itself will aid in maintenance of the historical file and will also make it easier to ascertain if any one SOP manual contains the current version of any individual SOP. Accessory documentation of effective dates (e.g., in the transmittal memo for the distribution of SOPs) is permissible but not recommended.

§ 58.83: Reagents and Solutions

> All reagents and solutions in the laboratory areas shall be labeled to indicate identity, titer or concentration, storage re-

quirements, and expiration date. Deteriorated or outdated reagents and solutions shall not be used.

Good laboratory technique has always included proper labeling of reagents and solutions. Many laboratories provide supplies of standard labels, which prompt laboratory personnel to include the four pieces of information mandated by the GLPs. "Identity" and "titer or concentration" present no problems. For "storage requirements" it is acceptable for laboratory SOPs to indicate that reagents and solutions may be stored at ambient room temperature unless otherwise indicated on the label. The standard label would then provide a space for "special storage conditions" (e.g., "refrigerate," "protect from light") The requirement to include an expiration date sometimes is resisted by laboratory personnel, especially for materials such as powder forms of histologic stains and crystalline sodium chloride. For such materials there is no known expiration date, and it is acceptable to indicate "NONE" or "N/A" (not applicable) on the label for expiration date. The laboratory must, however, be prepared to justify this designation. For other materials an expiration date should always be indicated on the label. The FDA has indicated that formal stability studies are not required to justify assigned expiration dates; it is sufficient to assign expiration dates based on literature references and/or laboratory experience.

The best guarantee that outdated reagents and solutions will not be used is a strictly enforced policy for discard of such materials, although that is not a GLP requirement. The GLPs require only that outdated materials not be used.

Official FDA enforcement policy requires adherence to GLP labeling requirements for all reagents and solutions in a laboratory in which GLP-regulated work is conducted even if some of those reagents and solutions are used for work that is not GLP-regulated. The FDA's concern is that reagents and solutions that are not adequately labeled, even if not intended for use in GLP-regulated studies, may have an adverse effect on laboratory work that is GLP-regulated.

§ 58.90: Animal Care

(a) There shall be standard operating procedures for the housing, feeding, handling, and care of animals.

This is simply a reiteration of the requirements of § 58.81.

(b) All newly received animals from outside sources shall be isolated and their health status shall be evaluated in accordance with acceptable veterinary medical practice.

Isolation is the separation of newly received animals from those already in the facility until the health of the newly received animals has been evaluated. Effective isolation minimizes the introduction of disease-causing agents into established animal colonies. It also allows time for the expression of clinical signs of disease, which will permit culling of animals before they are placed on study.

Quality control by the animal vendor and a knowledge of the history of the animals are acceptable parts of an institution's isolation procedures. This information may limit the isolation period for rodents to the time necessary for inspection upon arrival; however, all newly received animals should be allowed a stabilization period prior to their use [21].

Isolation may occur in the same room in which the study will be conducted; it is not necessary to provide separate, dedicated isolation areas. Laboratory personnel should solicit the expert advice of the veterinary staff in the establishment of isolation procedures.

(c) At the initiation of a nonclinical laboratory study, animals shall be free of any disease or condition that might interfere with the purpose or conduct of the study. If, during the course of the study, the animals contract such a disease or condition, the diseased animals shall be isolated, if necessary. These animals may be treated for disease or signs of disease provided that such treatment does not interfere with the study. The diagnosis, authorizations of treatment, description of treatment, and each date of treatment shall be documented and shall be retained.

Good science has always mandated the use of high-quality, disease-free animals to reduce extraneous factors that might complicate the interpretation of experimental results.

The GLPs permit the treatment of diseases or conditions that develop during the course of a study. Any animal so treated should be isolated from other animals if necessary to protect against adverse effects on a study. Laboratories may elect to euthanize diseased animals rather than provide treatment.

If a laboratory elects to treat diseased animals, the GLPs specify documentation requirements for such treatment. These documentation requirements are straightforward and consistent with accepted veterinary medical practice.

> (d) Warm-blooded animals, excluding suckling rodents, used in laboratory procedures that require manipulations and observations over an extended period of time or in studies that require the animals to be removed from and returned to their home cages for any reason (e.g., cage cleaning, treatment, etc.) shall receive appropriate identification. All information needed to specifically identify each animal within an animal-housing unit shall appear on the outside of that unit.

There is no perfect system for identification of animals. Tattoos and color codes frequently fade and may need to be redone after 3 to 5 months. Toe clips and ear punches are occasionally obliterated by self-mutilation or mutilation by cage mates. Ear tags and collars fall off and need to be replaced. Cage cards can be lost or destroyed. Whatever system or combination of systems of animal identification are selected by a laboratory, the shortcomings of the selected system(s) must be recognized, and procedures must be developed to address those shortcomings.

Identification other than a cage card is not required for short-term studies in which an animal is never taken from and returned to its cage during the course of a study. The preamble (¶ 35) to the GLP revisions of 1987 [4] also indicates that when animals are housed individually in cages, cage cards plus detailed animal-handling SOPs designed to prevent animal mix-ups will constitute an adequate animal identification system.

Any system of animal identification should provide an appropriate means for distinguishing one animal from all other animals housed in the same room. Each animal's identification only needs to be unique in the room in which it is housed; it need not be unique to all studies ever conducted with that species in the laboratory.

Identification of suckling rodents might lead to cannibalization by the mother, therefore FDA has exempted suckling rodents from the identification requirements of this section. There are other unique situations in which placing identifying features on an animal has the potential of jeopardizing the validity of the study. One such type of study is the guinea pig sensitization study, in which metal ear tags, plastic collars, or the dyes in tattoos and other color markings may themselves produce a sensitization response; ear punching may produce inflammation that could jeopardize test results; and toe clipping may lead to excessive bleeding. As previously mentioned, a laboratory may elect to identify such animals by cage card only. The animals must be singly housed, however, and animal-handling SOPs should provide specialized procedures for preventing animal mix-ups.

(e) Animals of different species shall be housed in separate rooms when necessary. Animals of the same species, but used in different studies, should not ordinarily be housed in the same room when inadvertent exposure to control or test articles or animal mixup could affect the outcome of either study. If such mixed housing is necessary, adequate differentiation by space and identification shall be made.

Physical separation of animals by species is generally recommended to prevent interspecies disease transmission and to reduce anxiety due to interspecies conflict. In some situations it might be appropriate to house different species of rodents in the same room, such as when they are to be used for tests of the same test article and have a similar health status or when special containment is provided within rooms (e.g., laminar flow cabinets or filtered or microisolation cages). It is not uncommon for animals from one supplier to harbor microbial agents not found in animals of the same species from another supplier, therefore intraspecies separation is advisable when animals obtained from multiple sources differ in microbiological status [21].

The best rule is only one species from a single supplier in any one room and only one study per room. If mixed housing is absolutely necessary, the laboratory must provide adequate differentiation by space and identification and must take steps to minimize the possibility for disease transmission or cross-contamination.

(f) Animal cages, racks and accessory equipment shall be cleaned and sanitized at appropriate intervals.

The National Institute of Health's (NIH) *Guide for the Care and Use of Laboratory Animals* [21] recommends that animal cages be sanitized before use, and further, that solid-bottom rodent cages be washed once or twice a week and cage racks at least monthly. It is recommended that wire-bottom cages and cages for all other animals be washed at least every 2 weeks. Water bottles, sipper tubes, stoppers, other watering equipment, and feeders should be washed once or twice a week.

The rinse cycle for washing all equipment should use water of at least 82.2°C (180°F), or higher for a period long enough to ensure destruction of vegetative pathogenic organisms. Chemical treatment is an alternative method of disinfection. If chemicals are used, equipment should be rinsed free of chemicals prior to use. Periodic microbiologic monitoring is useful to determine the efficacy of disinfection or sterilization procedures.

(g) Feed and water used for the animals shall be analyzed periodically to ensure that contaminants known to be ca-

> pable of interfering with the study and reasonably expected to be present in such feed or water are not present at levels above those specified in the protocol. Documentation of such analyses shall be maintained as raw data.

This GLP section was included as a result of FDA experience with toxicology studies of pentachlorophenol and diethylstilbestrol. In those studies the feeds used as carriers of the test article were found to contain varying quantities of pentachlorophenol and estrogenic activity. These contaminants invalidated the studies by producing erratic results.

Contaminant analysis of food and water for each and every study is not a requirement of § 58.90(g), nor is analysis for a laundry list of contaminants. What § 58.90(g) does require for every study is careful scientific consideration to determine whether or not there are any potential contaminants in the feed and water that are capable of interfering with test results. The study director and associated scientists from toxicology and other disciplines should consider each study in the light of its length, the expected toxicologic endpoints and pharmacologic activity of the test article, the test system, the route of administration, and other relevant factors to determine what contaminants could reasonably be expected to interfere. These considerations—coupled with scientific literature, experience, and anticipated levels of contamination–should be used to determine which, if any, contaminants should be controlled and analyzed. The FDA has said that it is unlikely that a blanket analysis conducted either by feed manufacturers or water authorities would be sufficient because such analyses would either provide data on contaminants that would not be expected to interfere or neglect to provide data for certain interfering contaminants.

Despite the foregoing, most labs rely on blanket analyses by feed manufacturers and water authorities, occasionally supplemented by analyses for a few additional contaminants also using a blanket approach (i.e., the same analyses for every study). It is likely that the type of scientific review expected by the FDA is simply not possible given the state of knowledge about test articles at the time safety studies are conducted.

Blanket analyses at least guard against the presence in the feed and water of known carcinogens (e.g., aflatoxin) that could interfere with the evaluation of a carcinogenicity study. Blanket analyses also assure that toxic materials (e.g., heavy metals, pesticides, Cominform bacteria) will not compromise the results of longer-term toxicity studies. Additional contaminant analyses should be conducted when the potential of interference by contaminants is known (e.g., tests for bivalent metal ions in the drinking water during the study of a tetracycline antibiotic and an analysis for estrogenic activity in the feed used during the study of an estrogen product).

The use of certified feeds for short-term studies is probably not justified unless a laboratory maintains only stocks of certified feeds to ensure that such feeds are used in longer-term studies. Such a policy also eliminates the need to maintain inventories of two types of feed for each species of animal.

When analyzing the animals' drinking water for possible interfering contaminants, representative water samples should be drawn at the point of use by the animals to detect any possible contamination of the water by the delivery system.

Most laboratories describe their blanket analyses for contaminants in SOPs, which provide a full listing of the contaminants analyzed and the acceptable levels for each. Study protocols in such cases merely make reference to the SOPs. If there is any analysis for contaminants not listed in the SOPs, the protocol should describe the additional contaminants and the acceptable levels for each.

> (h) Bedding used in animal cages or pens shall not interfere with the purpose or conduct of the study and shall be changed as often as necessary to keep the animals dry and clean.

Bedding should be absorbent, free of toxic chemicals or other substances that could injure animals or personnel, and of a type not readily eaten by animals. Bedding should be sufficient to keep animals dry between cage changes without coming into contact with watering tubes. Aromatic hydrocarbons from cedar and pine bedding materials can induce the biosynthesis of hepatic microsomal enzymes, therefore, such beddings are not appropriate for use in nonclinical laboratory studies.

Bedding can be purchased that is guaranteed to be free of potentially interfering contaminants. In the absence of such a guarantee, the laboratory may wish to consider its own periodic analysis of bedding for contaminants.

Bedding used in cages or pens should be changed as often as is required to keep the animals dry and clean. For small rodents (e.g., rats, mice, and hamsters) one to three bedding changes per week will generally suffice. For larger animals (e.g., dogs, cats, and nonhuman primates) bedding should be changed daily.

Soiled bedding should be emptied from cages and pans under conditions that minimize exposure of animals and personnel to aerosolized wastes.

> (i) If any pest control materials are used, the use shall be documented. Cleaning and pest control materials that interfere with the study shall not be used.

The most effective pest control program prevents entry of vermin into a facility by screening openings, sealing cracks, and eliminating breeding and refuge sites. With the exception, perhaps, of boric acid or drying substances (e.g., silica gel), there are few pest control materials that are free of serious toxic properties, therefore the best policy is one that prohibits the use of toxic pesticides in rooms in which animals are housed. If pest control materials are used in empty rooms, the room should not be used to house animals until the risk to animals has passed. This requires a knowledge of the degradation properties of the pesticide.

Application of pesticides must be recorded. The application must comply with federal, state, and local legal and regulatory requirements.

SUBPART F: TEST AND CONTROL ARTICLE

§ 58.105: Test and Control Article Characterization

(a) The identity, strength, purity, and composition or other characteristics which will appropriately define the test or control article shall be determined for each batch and shall be documented. Methods of synthesis, fabrication, or derivation of the test and control articles shall be documented by the sponsor or the testing facility. In those cases where marketed products are used as control articles, such products will be characterized by their labeling.

The definition of "appropriate" characterization of test and control articles will vary, depending on the stage of development of the articles. The amount of information on the first milligram quantity of material that is synthesized in the research laboratory will be much less than that available later in development when methods of synthesis have been scaled up to produce kilogram quantities. For test and control articles used in nonclinical laboratory studies, laboratory management should establish acceptable characteristics that are reasonably related to the stage of development.

Tests to characterize a test or control article as to its "identity" may be postponed until initial toxicology studies show a reasonable promise of the article's reaching the marketplace. The FDA has indicated, however, that information on "strength" and "purity" should be available prior to the use of the article in a nonclinical laboratory study.

Methods of synthesis, fabrication, or derivation as well as identity (if established), strength, and purity characteristics of the material must be documented. Copies of this documentation must be included with study records and must be available for FDA inspection. In the case of contract testing facilities in which for proprietary reasons the sponsor may not wish to release such information to the contract lab, the contract facility should have written assurance from the sponsor that such documentation exists.

Tests to establish the identity, strength, and purity of the test and control articles need not comply strictly with GLP requirements (e.g., protocol, QAU inspection requirements), but good documentation of analytical test results (usually in a laboratory notebook) and retention of raw data for such tests is a good practice. As the development process proceeds and the same material is used in both nonclinical and clinical studies, CGMP principles will apply to the production and characterization processes.

When marketed products are used as control articles, a copy of product labeling should be included with the study records.

> (b) The stability of each test or control article shall be determined by the testing facility or by the sponsor either: (1) Before study initiation, of (2) concomitantly according to written standard operating procedures, which provide for periodic analysis of each batch.

In most cases the stability of test articles will not be established before the initiation of a study. In such cases laboratory SOPs should describe a policy for periodic reanalysis of each batch of the test article. Analytical methods for reanalysis must be stability-indicating. The periodicity of the reanalysis is left to the discretion of the laboratory. Generally analyses are conducted at 3- to 6-month intervals during the period of test article use. In establishing the analysis interval the laboratory will want to weight the risk of loss of a study because of test or control article instability against the costs of the periodic reanalyses.

The periodic stability reanalyses must be conducted in full compliance with the GLP regulations.

> (c) Each storage container for a test or control article shall be labeled by name, chemical abstract number or code number, batch number, expiration data, if any, and, where appropriate, storage conditions necessary to maintain the identity, strength, purity, and composition of the test or control article. Storage containers shall be as-

> signed to a particular test article for the duration of the study.

Labeling requirements in § 58.105(c) are not controversial and are the minimum to ensure against mix-up of test or control articles. The Expiration date needs to be included on the label only if one has been established. Some laboratories include a retest date on the label as a reminder of the need for periodic stability analyses. Only special storage conditions (e.g., "refrigerate," "protect from light," "protect from freezing") need to be included on the label.

In the preamble (¶ 38) to the 1987 GLP revisions [4], FDA declined to eliminate the storage container provision in § 58.105(c). Dr. Paul Lepore has indicated that ¶ 38 referred only to the original storage container. According to the scenario envisioned by FDA, a lot of test or control article is selected for testing, characterized, and placed in a properly labeled storage container. This storage container must be retained for the duration of the test. Aliquots or samples of test article may be removed from this storage container and placed in intermediate "working" containers that are also properly labeled. However, these "working" containers need not be retained.

> (d) For studies of more than 4 weeks' duration, reserve samples from each batch of test and control articles shall be retained for the period of time provided by § 58.195. (Collection of information requirements approved by the Office of Management and Budget under number 0910–0203)

The FDA has indicated that "study initiation date" (defined in § 58.3(o)) and "study completion date" (defined in § 58.3(p)) are administrative dates and should not be used to determine whether or not a study is "of more than 4 weeks' duration." Instead, terms-of-the-art (e.g., 14-day acute study, 28-day repeated dose study, etc.) will determine whether reserve samples are required under § 58.105(d).

Reserve sample size should be at least twice the quantity necessary to perform all tests to determine whether the test or control article meets its established specifications for identity, strength, quality, purity, and stability. By retaining twice the quantity necessary to perform all tests, the laboratory will be able to supply a sample to the FDA, if requested, and still retain sufficient material to conduct its own tests.

§ 58.107: Test and Control Article Handling

Procedures shall be established for a system for the handling of the test and control articles to ensure that

(a) There is proper storage.
(b) Distribution is made in a manner designed to preclude the possibility of contamination, deterioration, or damage.
(c) Proper identification is maintained throughout the distribution process.
(d) The receipt and distribution of each batch is documented. Such documentation shall include the date and quantity of each batch distributed or returned.

The general goals of § 58.107 are to maintain the integrity of and to provide accountability for the test and control articles throughout the period of use.

Integrity is maintained by ensuring that all containers of the articles are labeled properly, by storing all supplies of the articles in conformance with their labeling, and by ensuring that the articles are distributed, handled, and used in a manner that precludes the possibility of contamination, deterioration, or damage.

The accountability provisions of § 58.107(d) are met by records showing the date and quantity of test and control articles distributed from central stores for use in a study or series of studies and the date and amount of material returned to central stores at the end of a study or amount of material returned to central stores at the end of a study or series of studies. To this should be added a system for documenting the date and quantity for each use of a test or control article during the course of each study. A running inventory of test and control articles is not required but does provide an easy mechanism for periodically verifying the accuracy of test and control article usage.

§ 58.113: Mixtures of Articles with Carriers

(a) For each test or control article that is mixed with a carrier, tests by appropriate analytical methods shall be conducted:
   (1) To determine the uniformity of the mixture and to determine, periodically, the concentration of the test or control article in the mixture.
   (2) To determine the stability of the test and control articles in the mixture as required by the conditions of the study either (i) before study initiation, or (ii) concomitantly according to written standard operating procedures which provide for periodic analysis of the test and control articles in the mixture.

The requirements of § 58.113(a) substantially changed the state of the art for the conduct of nonclinical laboratory studies. Prior to the promulgation of GLP regulations, analytical tests to establish the homogeneity and stability of article/carrier mixtures were not routine, nor were tests to determine the concentration of test and control articles in the mixtures used to deliver test and control articles to test systems.

There is no exemption from § 58.113(a) for short-term studies. If a study meets the definition of a nonclinical laboratory study all analytical requirements apply.

If a test or control article is administered in solution, homogeneity (uniformity) tests need not be conducted. For nonsolutions (e.g., suspensions and mixtures with diet), once the uniformity has been established for a given set of mixing conditions, it is not necessary to establish the uniformity of each subsequent batch that is mixed according to the same specifications. In taking samples for homogeneity testing, one must ensure that the samples are truly representative of the batch and that the total number of samples is adequate to prove uniformity. Typically samples are drawn from the top, middle, and bottom of the batch or according to a random sampling schedule. The number of samples from any one batch usually ranges from 6 to 9.

Stability of the article/carrier mixture can be established in conjunction with the homogeneity assays of nonsolutions. Separate stability tests will, of course, be required for solutions. Formal stability trials sufficient to show long-term stability of the mixtures are not required; rather, stability should be established for a period that encompasses the period of use of the article/carrier mixture. Period of use should be defined as whichever of the following two time periods is longer, the time between preparation of the mixture and final administration of that mixture to the test system, or the time between preparation of the mixture and the analysis of the mixture as required by § 58.113(a)(2). Often the period between preparation and analysis may be longer than the period between preparation and last administration to the test system.

Homogeneity and stability assays may be conducted before a study begins or may be conducted concurrently with the study. If the latter, poor assay results may, or course, result in invalidation of the study.

There are no established guidelines with regard to the frequency of periodic concentration assays. Some laboratories randomly select a sample from one concentration of article/carrier mixture per study per week. Other laboratories conduct an analysis of all concentrations of article/carrier mixtures on a monthly or quarterly basis.

When article/carrier mixtures are prepared by serial dilution of the highest concentration, FDA has suggested that it would be appropriate to perform an assay on the lowest concentration because this would confirm the accuracy of the dilution process. This is not a GLP requirement, however, and there is no prohibition on the analysis of any of the other concentrations. Analytical methods may not be sensitive enough for valid assays of the lowest concentration.

Although some laboratories do not use any article/carrier mixture until satisfactory analytical results are obtained from a concentration assay of the mixture, this is not a GLP requirement. The concentration assays provide periodic assurance that test systems are being exposed to the amounts and types of test and control articles that are specified in the protocol, therefore the results of the periodic concentration assays must be reviewed critically and promptly. Analytical results outside a pre-established acceptable range (as defined by laboratory SOPs) will require follow-up. Follow-up should attempt to determine the cause of poor analytical results (e.g., improper preparation of the article/carrier mixture, sample mix-up, poor analytical technique, equipment malfunction). Corrective action should then be provided as necessary. Usually analytical results in excess of 10% above or below expected values will require follow-up.

Tests to establish the stability and homogeneity of article/carrier mixtures as well as the periodic concentration analyses of the mixtures must be conducted in full compliance with the GLP regulations.

(b) (Reserved)

(c) Where any of the components of the test or control article carrier mixture has an expiration date, that date shall be clearly shown on the container. If more than one component has an expiration date, the earliest date shall be shown.

A reasonable interpretation of § 58.113(c) should not require expiration dating of containers of article/carrier mixtures when the mixtures will be used on the date of preparation unless a component of the mixture has an extremely short (e.g., less than 8 hr) period of stability. This section does not require that an expiration date appear on feeders that are filled with article/diet mixtures on the date the mixture is prepared and are presented to the test animals on that same day.

SUBPART G: PROTOCOL FOR AND CONDUCT OF A NONCLINICAL LABORATORY STUDY

§ 58.120: Protocol

(a) Each study shall have an approved written protocol that clearly indicates the objectives and all methods for the conduct of the study. The protocol shall contain, as applicable, the following information:

The requirement to indicate "all methods for the conduct of the study" does not mean that all laboratory SOPs must be reiterated in the protocol; it is sufficient if the protocol indicates "what" will be done and "when" it will be done. Laboratory SOPs describe "how" each study activity is to be performed. If exceptions from SOPs will apply for the study, then those exceptions should be described in the protocol. The FDA has indicated that the protocol should list the SOPs used in a particular study, but the author suggests that a simple stipulation in the protocol that "the study will be conducted in accordance with current standard operating procedures" is sufficient. Listing each SOP in the protocol could cause problems if SOP identifying numbers or titles change during the course of a study.

All of the following items, if relevant, must be included in the protocol:

1. A descriptive title and statement of the purpose of the study.
2. Identification of the test and control articles by name, chemical abstract number, or code number.
3. The name of the sponsor and the name and address of the testing facility at which the study is being conducted.
4. The number, body weight range, sex, source of supply, species, strain, substrain, and age of the test system.
5. The procedure for identification of the test system.
6. A description of the experimental design, including the methods for the control of bias.
7. A description and/or identification of the diet used in the study as well as solvents, emulsifiers, and/or other materials used to solubilize or suspend the test or control articles before mixing with the carrier. The description shall include specifications for ac-

ceptable levels of contaminants that are reasonably expected to be present in the dietary materials and are known to be capable of interfering with the purpose or conduct of the study if present at levels greater than established by the specifications.

If a laboratory conducts a blanket analysis for contaminants, the protocol can make reference to a description of those analyses in laboratory SOPs. Any additional analyses that are specific to the study should be described in the protocol.

8. Each dosage level, expressed in mg per kg of body weight or other appropriate units, of the test or control article to be administered and the method and frequency of administration
9. The type and frequency of tests, analyses, and measurements to be made
10. The records to be maintained
11. The date of approval of the protocol by the sponsor and the dated signature of the study director
12. A statement of the proposed statistical methods to be used

It is important to describe statistical methods in the protocol. This will avoid suspicions that statistical methods were selected after study data were available and that selection was based on a desired end result.

A protocol is required for each nonclinical laboratory study. Usually a single protocol will cover only one experiment with a single test article in a single type of test system. It is permissible, however, to conduct several experiments using the same test article under a single comprehensive protocol. It is also permissible to study several test articles concurrently using a single common procedure under one protocol.

The intent of § 58.120 is to provide all study personnel with clear directions as to the objectives of a study and all operations needed to fulfill those objectives, therefore even though not required by § 58.120, it is important that all personnel involved with a study have access to a copy of the protocol and all amendments. Such access best assures that study procedures will be done as and when intended.

> (b) All changes in or revisions of an approved protocol and the reasons therefore shall be documented, signed by the study director, dated, and maintained with the protocol. (Collection of information requirements approved by the Office of Management and Budget under number 0919–0203)

Documentation of protocol changes or revisions and the reason for them is best accomplished by issuing formal protocol amendments, which must be dated and signed by the study director and should be attached to the front of all copies of the protocol. Such attachments immediately alerts study personnel to protocol changes and help prevent study personnel from overlooking amendments that are "hidden" at the back of the protocol.

If deviations from a protocol are intended to be permanent, a protocol amendment should be issued to document the change. If a deviation from the protocol is an error, the deviation should be promptly corrected and should be documented in the study records and described in the final report.

To the extent possible, protocol amendments should be prospective; that is, issued and distributed before the change is intended to occur. In some circumstances (e.g., an emergency decision to lower test article dose levels in a chronic study because of an unexpected toxic response to protocol-specified doses or a decision to collect additional tissue specimens where that decision is made on the basis of findings during the course of an autopsy) prospective distribution of a protocol amendment may not be possible. In such cases, a protocol amendment should be issued as soon as possible.

The question frequently arises as to what date should appear on a protocol amendment. The author is of the opinion that an effective date, whether prospective or retrospective, should be included. An effective date will alert personnel to the date when the amendment goes into effect and will also provide a historical record of the time period during which the amendment was in effect. It may also be helpful to include an issue date.

§ 58.130: Conduct of a Nonclinical Laboratory Study

(a) The nonclinical laboratory study shall be conducted in accordance with the protocol.
(b) The test systems shall be monitored in conformity with the protocol.

Sections 8.130(a) and (b) should not be regarded as a straitjacket that prevents scientifically justified changes in research as a study progresses. Any changes in the research can occur as long as they are properly documented in the form of protocol amendments. There is no limit on the number of amendments.

(c) Specimens shall be identified by test system, study, nature, and date of collection. This information shall be located on the specimen container or shall accompany the specimen in a manner that precludes error in the recording and storage of data.

The proper identification of specimens is of obvious importance to the validity of a study. Because of the size or nature of the material, some types of specimens (e.g., paraffin blocks and microscopic slides) do not lend themselves to labeling for all the items listed in § 58.130(c). In such cases the use of an alternative identification (e.g., accession numbers) is acceptable as long as the alternate identification can be translated into the required information.

In some instances, failure to include the "nature" of the specimen will not be contrary to the intent of the regulations. For example, a microscopic slide that contains liver tissue need not have "liver" written on the slide since the end user, the diagnosing pathologist, will not need the label to identify the tissue as liver. On the other hand, sections of tumor from a multiple tumor-bearing animal should be clearly labeled to indicate from which tumor the sections were taken.

"Shall accompany the specimen" need not be strictly interpreted in the case of archive material. For example, a specimen labeled with an accession number can be stored in the specimen archives while the document that translates the accession number into the additional label information is stored in a separate document archive. As long as both the specimen and the associated document are readily retrievable, the intent of the regulations is met.

(d) Records of gross findings for a specimen from postmortem observations should be available to a pathologist when examining that specimen histopathologically.

To better ensure that the pathologist will be prompted to provide microscopic follow-up to all grossly observed lesions, it is important that information on gross findings be available to the diagnosing pathologist.

There may be occasions when study design requires that information on gross findings be withheld from the diagnosing pathologist (e.g., in the case of totally blinded slide reading). This is permissible, but FDA does not believe that "blinding" is a preferred practice in histopathologic evaluation.

(e) All data generated during the conduct of a nonclinical laboratory study, except those that are generated by automated data collection systems, shall be recorded directly, promptly, and legibly in ink. All data entries shall be dated on the day of entry and signed or initialed by the person entering the data. Any change in entries shall be made so as not to obscure the original entry, shall indicate the reason for such change, and shall be dated and signed or identified at the time of the change. In automated data collection systems, the individual responsible for direct data input shall be identified at the time of data input. Any change in automated data entries shall be made so as not to obscure the original entry, shall indicate the reason for change, shall be dated, and the responsible individual shall be identified.

(Collection of information requirements approved by the Office of Management and Budget under number 0919–0203)

All data must be recorded promptly (defined by Webster as "immediately"). Hand-recorded data must be recorded in ink (to prevent improper erasures and corrections).

A signature (or initials) and date are not required for every individual piece of data. It is sufficient, for example, to provide one signature (or initials) and date for all data collected during a single data collection session. The purpose of the signature or initials is to provide accountability for the data.

The CGMP regulations [10] require certain activities (e.g., charge-in of components) to be performed by one individual and witnessed and verified by a second individual. There is no similar requirement in the GLP regulations, but some laboratories voluntarily elect to have certain critical operations (e.g., test article weighings) witnessed and verified by a second individual.

With the exception of automated data collection systems, all changes in data should be made by drawing a single line through the data being changed, recording the corrected or changed information and the date of change, and indicating a reason for the change. The person making the change should be identified by signature or initials. The explanation of the change need not be elaborate. For example, "number transposition" or "entered in wrong column" can suffice as an explanation. Simply indicating "error" is seldom an adequate explanation. A coded system (e.g., number or letter) of recording the reasons for data changes is acceptable if the code is translated on the data form or in laboratory SOPs. The need to document reasons for changes in data must be constantly reinforced with study personnel.

Special rules apply in the case of automated data collection systems; the person responsible for data collection must be identified at the time of data input. Changes in automated data entries must be made in such a way that the original entry is saved, and the person responsible for making the change must be identified. The other requirements for data changes, recording the reason for the change and the date the change was made, also apply to automated data collection systems. The audit trail for changes in automated data entries may be recorded on paper or on computer.

SUBPARTS H–I: (RESERVED)

SUBPART J: RECORDS AND REPORTS

§ 58.185: Reporting of Nonclinical Laboratory Study Results

(a) A final report shall be prepared for each nonclinical laboratory study and shall include, but not necessarily be limited to, the following:

With the exception of the second sentence of item 7, all of the following topics must be addressed in the final report. Unlike § 58.120(a), the words "as applicable" do not appear in § 58.185(a), thus for example, the report must address the issue of statistical analysis even if no statistical analysis was required or done.

(1) Name and address of the facility performing the study and the dates on which the study was initiated and completed.

The FDA requires the name and address of the testing facility to appear in the report so that when the report is submitted in support of a research or marketing permit the laboratory can be added to FDA's inventory of laboratories that are scheduled for GLP inspection. The name and address may also be used by FDA to establish the site for any directed audit of the report.

(2) Objectives and procedures stated in the approved protocol, including any changes in the original protocol.
(3) Statistical methods employed for analyzing the data.
(4) The test and control articles identified by name, chemical abstracts number or code number, strength, purity, and composition or other appropriate characteristics.

(5) Stability of the test and control articles under the conditions of administration.

The "stability . . . under the conditions of administration" will in most cases be the stability of the article/carrier mixtures determined under § 58.113(a)(2). If a drug is administered as a powder (e.g., by capsule), the stability of the bulk drug determined under § 58,105(b) will be reported.

(6) A description of the methods used.
(7) A description of the test system used. Where applicable, the final report shall include the number of animals used, sex, body weight range, source of supply, species, strain and substrain, age, and procedure used for identification.
(8) A description of the dosage, dosage regimen, route of administration, and duration.
(9) A description of all circumstances that may have affected the quality or integrity of the data.

Under § 58.33(c) the study director is responsible for documenting all circumstances that may affect the quality and integrity of the study. Such circumstances must be described in the final report.

(10) The name of the study director, the names of other scientists or professionals, and the names of all supervisory personnel, involved in the study.

Only the names of study personnel need be listed in the report. The signatures required are those of the study director and those individuals described in § 58.185(a)(12). A laboratory is permitted some discretion in the listing of names. The names of technicians and animal-care workers need not be listed. The list of names is usually limited to senior scientific or supervisory staff.

(11) A description of the transformations, calculations, or operations performed on the data, a summary and analysis of the data, and a statement of the conclusions drawn from the analysis.
(12) The signed and dated reports of each of the individual scientists or other professionals involved in the study.

In the preamble ({{paragraph}} 48a) to the 1987 GLP revisions [4], FDA rejected a request to modify § 58.185(a)(12) to permit combined reports signed by the principal scientists (e.g., clinical veterinarian, clinical pathologist, histopathologist). The FDA stated that each individual scientist involved in a study must be accountable for reporting data, information, and views within his or her designated area of responsibility and that combined reports would obscure the individual's accountability for accurate reporting.

Prior to publication of the 1987 GLP revisions, many laboratories prepared combined reports, and the author knows of no instance in which the FDA rejected a study for failure to provide signed and dated reports from each of the scientists or other professionals involved in the study. For such laboratories it is probably advisable to reconsider prior policy on report preparation. The intent of the regulation (to provide accountability) can be met with the format of a combined report, but with an indication on the signature page of the portion of the report prepared by each signatory.

It is customary to append the signed reports of consultants (e.g., consulting ophthalmologists, consulting pathologists) to the reports submitted to FDA.

(13) The location where all specimens, raw data, and the final report are to be stored.

In most cases these materials will be stored in the archives of the testing facility, and the report will so indicate. In the case of contract safety testing, however, a sponsor will sometimes ask that raw data, documentation, and specimens be sent to the sponsor for storage in the sponsor's archives. In other cases a laboratory may store some or part of the archival material at an off-site location. In either case the final report should reference the actual storage site(s).

(14) The statement prepared and signed by the quality assurance unit as described in § 58.35(b)(7).

(b) The final report shall be signed and dated by the study director.

(c) Corrections or additions to a final report shall be in the form of an amendment by the study director. The amendment shall clearly identify that part of the final report that is being added to or corrected and the reasons for the correction or addition, and shall be signed and dated by the person responsible.

A report becomes "final" when it is signed by the study director. Any changes in the report after it is signed by the study director must be in the form of an amendment that meets the requirements of

§ 58.185(c). To avoid the necessity for many report amendments, the report should not be signed by the study director until it has been reviewed by the scientists involved in the study and has been audited by the QAU and after all changes and corrections occasioned by that review and audit have been made.

The purpose of § 58.185(c) is to guard against inappropriate or unwarranted changes being made in the report without the knowledge and concurrence of the study director.

§ 58.190: Storage and Retrieval of Records and Data

(a) All raw data, documentation, protocols, final reports and specimens (except those specimens obtained from mutagenicity tests and wet specimens of blood, urine, feces, and biological fluids) generated as a result of a nonclinical laboratory study shall be retained.

(b) There shall be archives for orderly storage and expedient retrieval of all raw data, documentation, protocols, specimens, and interim and final reports. Conditions of storage shall minimize deterioration of the documents or specimens in accordance with the requirements for the time period of their retention and the nature of the documents or specimens. A testing facility may contract with commercial archives to provide a repository for all material to be retained. Raw data and specimens may be retained elsewhere provided that the archives have specific reference to those other locations.

Any laboratory that conducts nonclinical laboratory studies must provide dedicated space for the storage of raw data, documentation, protocols, specimens, and interim and final reports from completed studies. The laboratory must have an orderly system for storing such material, and that system must provide an expedient method for retrieving of archived materials (e.g., on the request of an FDA inspector).

Storage conditions (e.g., temperature, humidity) in the archives should be reasonably related to the nature of the stored documents, specimens, and samples. For example, wet tissues and paraffin blocks should be protected against extremes of high temperature, paper documents should not be subjected to long periods of high humidity, and reserve samples of test and control articles should be stored in accordance with label requirements. The FDA has indicated that "heroic" measures need not be taken to preserve materials in the archives, but storage conditions that foster accelerated deterioration should be avoided. Storage conditions should be monitored so that deviations from proper storage conditions can be promptly rectified.

If an off-site area is used to house the archives, whether owned or rented by the testing facility or operated by a commercial archival service, the on-site archives must contain specific reference to the materials that are stored off-site and the location of the alternate storage site(s).

(c) An individual shall be identified as responsible for the archives.

Similar to the requirements for a study director, a nonclinical testing laboratory must designate a single individual to be responsible for the archives. It is permissible to designate an alternate archivist to serve in the absence of the designated archivist.

(d) Only authorized personnel shall enter the archives.

Laboratory SOPs should define the personnel who may enter the archives. This need not be a list of names of individuals, but should provide adequate guidance to archive personnel as to who may enter the archives. Many laboratories allow only archive personnel to enter the archives but allow authorized personnel to check out archive material. If materials are removed from the archives for any reason, a record should be kept of what is removed and by whom. Follow-up should be provided by archive personnel to ensure prompt return of materials to the archives.

(e) Material retained or referred to in the archives shall be indexed to permit expedient retrieval.
(Collection of information requirements approved by the Office of Management and Budget under number 0910–0203)

Any indexing system for material in the archives is acceptable as long as the system permits rapid retrieval of archived materials.

§ 58.195: Retention of Records

(a) Record retention requirements set forth in this section do not supersede the record retention requirements of any other regulations in this chapter.

If the record retention requirements of § 58.195 are inconsistent with those of any other part of 21 CFR, the other parts of 21 CFR will take precedence.

(b) Except as provided in paragraph (c) of this section, documentation records, raw data and specimens pertaining to a nonclinical laboratory study and required to be made by this part shall be retained in the archive(s) for whichever of the following periods is shortest:
   (1) A period of at least 2 years following the date on which an application for a research or marketing permit, in support of which the results of the nonclinical laboratory study were submitted, is approved by the Food and Drug Administration. This requirement does not apply to studies supporting investigational new drug applications (IND's) or applications for investigational device exemptions

(IDE's), records of which shall be governed by the provisions of paragraph (b)(2) of this section.

(2) A period of at least 5 years following the date on which the results of the nonclinical laboratory study are submitted to the Food and Drug Administration in support of an application for a research or marketing permit.

(3) In other situations (e.g., where the nonclinical laboratory study does not result in the submission of the study in support of an application for a research or marketing permit), a period of at least 2 years following the date on which the study is completed, terminated, or discontinued.

Records, raw data, and specimens from a nonclinical laboratory study must be retained for whichever of the three time periods indicated above is shortest. An exception is made for those nonclinical laboratory studies that support an application for an IND or an IDE, for which records must be retained for a minimum of 5 years after the results of those studies are submitted to FDA.

Most companies take a more conservative approach and retain documents, microscopic slides, and paraffin blocks indefinitely. Materials that take up more storage space, such as wet tissues, are generally the first materials to be discarded. Paper documents may be discarded at any time if they have been converted to microfilm or microfiche.

(c) Wet specimens (except those specimens obtained from mutagenicity tests and wet specimens of blood, urine, feces, and biological fluids), samples of test or control articles, and specially prepared material which are relatively fragile and differ markedly in stability and quality during storage, shall be retained only as long as the quality of the preparation affords evaluation. In no case shall retention be required for longer periods than those set forth in paragraphs (a) and (b) of this section.

If a laboratory elects to discard fragile materials before the expiration of the applicable time period of § 58.195(b), the date of discard and the justification for discard should be recorded, and the documentation should be retained in the archives.

Examples of "specially prepared material" were listed in the GLP regulations prior to the 1987 revisions. These included histochemical, electron microscopic, blood mounts, and teratological preparations. These examples are illustrative and not comprehensive.

> (d) The master schedule sheet, copies of protocols, and records of quality assurance inspections, as required by § 58.35(c) shall be maintained by the quality assurance unit as an easily accessible system of records for the period of time specified in paragraphs (a) and (b) of this section.

The records and documents required to be maintained by the QAU are also subject to the record retention requirements of § 58.195(b). Spokesmen for the FDA have stated on occasion that these QA records should be stored in the archives described in § 58.190(b). This is an option that can be considered by the QAU, but there is no stipulated requirement in the GLP regulations for such storage. In fact, it could be argued that the requirement of § 58.35(a) for the QA function to be independent of nonclinical laboratory study personnel militates against storage of QA records in the archives.

> (e) Summaries of training and experience and job descriptions required to be maintained by § 58.29(b) may be retained along with all other testing facility employment records for the length of time specified in paragraphs (a) and (b) of this section.

Rather than storing summaries of training and experience and job descriptions in the GLP archives, a laboratory may elect to store such records together with other employment records (e.g., in the personnel department). If such alternative storage of these records is elected, care should be taken that the personnel responsible for the alternate records storage are aware of GLP record retention requirements. Before electing such alternate storage, a system should be established to preserve the confidentiality of the personnel records (other than summaries of training and experience and job descriptions) at the time of FDA or QA inspections.

(f) Records and reports of the maintenance and calibration and inspection of equipment, as required by § 58.63(b) and (c), shall be retained for the length of time specified in paragraph (b) of this section.

Records of the maintenance, calibration, and inspection of equipment are also subject to the record retention requirements of the regulations. Often a facility has its own metrology group or contracts with an outside group to handle maintenance and calibration of equipment. In such cases the records of these activities may include records for equipment used in both GLP and non-GLP archives. This is not contrary to GLP requirements as long as the regular GLP archives makes reference to the alternate storage place and as long as the alternate storage meets the GLP requirements for secure and orderly storage, expedient retrieval of records, limited access to the records storage area, and responsibility for storage under the control of a single individual.

(g) Records required by this part may be retained either as original records or as true copies such as photocopies, microfilm, microfiche, or other accurate reproductions of the original records.

The 1987 GLP revisions added § 58.195(g) to re-emphasize long-standing FDA policy that a laboratory may retain either original records or accurate reproductions of them. It should be noted that magnetic media may qualify as either original records or accurate reproductions of same.

(h) If a facility conducting nonclinical testing goes out of business, all raw data, documentation, and other material specified in this section shall be transferred to the archives of the sponsor of the study. The Food and Drug Administration shall be notified in writing of such a transfer.

Going out of business is often a sudden and unplanned event for a laboratory. Under such circumstances, personnel from the lab ceasing operations may not show proper concern for complying with regulatory requirements. The party with the greatest stake in preserving the records of a study, namely the sponsor, may therefore have to assume responsibility for the preservation and transfer of the records to the sponsor's location and for notifying FDA of the transfer.

There was an instance in which a laboratory that had conducted a number of studies for EPA regulatory purposes went out of business and the records relating to studies the laboratory had conducted were lost. In this case EPA required many of the studies to be repeated. The lesson to be learned from this experience is that a sponsor should be very careful in the selection of contract facilities and should periodically check with the contract lab to ensure that the laboratory continues to operate and that study records continue to be maintained. Some sponsors obtain the specimens and/or originals or copies of all raw data for contracted studies for storage in their own archives to protect against the loss of raw data at the contract laboratory.

SUBPART K: DISQUALIFICATION OF TESTING FACILITIES

§ 58.200: Purpose

(a) The purposes of disqualification are: (1) To permit the exclusion from consideration of completed studies that were conducted by a testing facility which has failed to comply with the requirements of the good laboratory practice regulations until it can be adequately demonstrated that such noncompliance did not occur during, or did not affect the validity or acceptability of data generated by, a particular study; and (2) to exclude from consideration all studies completed after the date of disqualification until the facility can satisfy the Commissioner that it will conduct studies in compliance with such regulations.

Disqualification is the most severe penalty that FDA can apply for failure to comply with GLP requirements. If a laboratory is disqualified, the completed or future studies conducted by that laboratory may not be accepted by FDA in support of an application for a research or marketing permit. It is even possible for prior FDA approval of a marketed product to be withdrawn if that approval was based in part on the study or studies conducted by a disqualified laboratory.

(b) If a sponsor is actively pursuing a research or marketing permit for a test article and if a disqualified laboratory has conducted a nonclinical laboratory study on that test article, the sponsor is still obligated to submit the results of such a study to the FDA. As indicated in paragraph (a), FDA will not consider the results of that study in support of the research or marketing permit, but FDA may use the results of the study in reaching a conclusion that the research or marketing permit should not be approved. Thus the results of a study conducted by a disqualified laboratory can never work to the sponsor's advantage but may work to the sponsor's disadvantage.

§ 58.202: Grounds for Disqualification

The commissioner may disqualify a testing facility upon finding all of the following:

(a) The testing facility failed to comply with one or more of the regulations set forth in this part (or any other regulations regarding such facilities in this chapter);
(b) The noncompliance adversely affected the validity of the nonclinical laboratory studies; and
(c) Other lesser regulatory actions (e.g., warnings or rejection of individual studies) have not been or will probably not be adequate to achieve compliance with the good laboratory practice regulations.

It is important to note that the FDA must find all three conditions, as indicated in paragraphs (a), (b), and (c) above, before it can disqualify a laboratory.

Since the effective date of the GLP regulations, no laboratory has been disqualified. The FDA has, however, issued many warnings and has rejected individual studies for reasons of GLP noncompliance.

There are instances in which laboratories have gone out of business "voluntarily" because they lack the desire or ability to comply with GLP requirements.

With the exception of § 58.217, the balance of subpart K describes the legal and administrative procedures that govern the disqualification process. The remaining sections of subpart K are reprinted for the sake of completeness but will not be commented on, with the exception of § 58.217.

§ 58.204: Notice of and Opportunity for Hearing on Proposed Disqualification

(a) Whenever the Commissioner has information indicating that grounds exist under § 58.202 which in his opinion justify disqualification of a testing facility, he may issue to the testing facility a written notice proposing that the facility be disqualified.

(b) A hearing on the disqualification shall be conducted in accordance with the requirements for a regulatory hearing set forth in Part 16.

§ 58.206 Final Order on Disqualification

(a) If the Commissioner, after the regulatory hearing, or after the time for requesting a hearing expires without a request being made, upon an evaluation of the administrative record of the disqualification proceeding, makes the findings required in § 58.202, he shall issue a final order disqualifying the facility. Such order shall include a statement of the basis for that determination. Upon issuing a final order, the Commissioner shall notify (with a copy of the order) the testing facility of the action.

(b) If the Commissioner, after a regulatory hearing or after the time for requesting a hearing expires without a request being made, upon an evaluation of the administrative record of the disqualification proceeding, does not make the findings required in § 58.202, he shall issue a final order terminating the disqualification proceeding. Such order shall include a statement of the basis for that determination. Upon issuing a final order the Commissioner shall notify the testing facility and provide a copy of the order.

§ 58.210: Actions upon Disqualification

(a) Once a testing facility has been disqualified, each application for a research or marketing permit, whether approved or not, containing or relying upon any nonclinical laboratory study conducted by the disqualified testing facility may be examined to determine whether such study was or would be essential to a decision. If it is determined that a study was or would be essential, the Food and Drug Administration shall also determine whether the study is acceptable, notwithstanding the disqualification of the facility. Any study done by a testing facility before or after disqualification may be presumed to be unacceptable, and the person relying on the study may be required to establish that the study was not affected by the circumstances that led to the disqualification, e.g., by submitting validating information. If the study is then determined to be unacceptable, such data such [sic] be eliminated from consideration in support of the application; and such elimination may serve as new information justifying the termination or withdrawal of approval of the application.

(b) No nonclinical laboratory study begun by a testing facility after the date of the facility's disqualification shall be considered in support of any application for a research or marketing permit, unless the facility has been reinstated under § 58.219. The determination that a study may not be considered in support of an application for a research or marketing permit does not, however, relieve the applicant for such a permit of any obligation under any other applicable regulation to submit the results of the study to the Food and Drug Administration.

§ 58.213: Public Disclosure of Information Regarding Disqualification

(a) Upon issuance of a final order disqualifying a testing facility under § 58.206(a), the Commissioner may notify all or any interested persons. Such notice may be given at the discretion of the Commissioner whenever he believes that such disclosure would further the public interest or would promote compliance with the good laboratory practice regulations set forth in this part. Such notice, if given, shall include a copy of the final order issued under § 58.206(a) and shall state that the disqualification constitutes a determination by the Food and Drug Administration that nonclinical laboratory studies performed by the facility will not be considered by the Food and Drug Administration in support of any application for a research or marketing permit. If such notice is sent to another Federal Government agency, the Food and Drug Administration will recommend that the agency also consider whether or not it should accept nonclinical laboratory studies preformed by the testing facility. If such notice is sent to any other person, it shall state that it is given because of the relationship between the testing facility and the person being notified and that the Food and Drug Administration is not advising or recommending that any action be taken by the person notified.

(b) A determination that a testing facility has been disqualified and the administrative record regarding such determination are disclosable to the public under Part 20.

§ 58.215: Alternative or Additional Actions to Disqualification

(a) Disqualification of a testing facility under this subpart is independent of, and neither in lieu of nor a precondition to, other proceedings or actions authorized by the act. The Food and Drug Administration may, at any time, institute against a testing facility and/or against the sponsor of a nonclinical laboratory study that has been submitted to the Food and Drug Administration any appropriate judicial proceedings (civil or criminal) and any other appropriate regulatory action, in addition to or in lieu of, and prior to, simultaneously with, or subsequent to, disqualification. The Food and Drug Administration may also refer the matter to another Federal, State, or local government law enforcement or regulatory agency for such action as that agency deems appropriate.

(b) The Food and Drug Administration may refuse to consider any particular nonclinical laboratory study in support of an application for a research or marketing permit, if it finds that the study was not conducted in accordance with the good laboratory practice regulations set forth in this part, without disqualifying the testing facility that conducted the study or undertaking other regulatory action.

§ 58.217: Suspension or Termination of a Testing Facility by a Sponsor

Termination of a testing facility by a sponsor is independent of, and neither in lieu of nor a precondition to, proceedings or actions authorized by this subpart. If a sponsor terminates or suspends a testing facility from further participation in a nonclinical laboratory study that is being conducted as part of any application for a research or marketing permit that has been submitted to any Center of the Food and Drug Administration (whether approved or not), it shall notify that Center in writing within 15 working days of the action; the notice shall include a statement of the reasons for such action. Suspension or termination of a testing facility by a sponsor does not relieve it of any obligation under any other applicable regulation to submit the results of the study to the Food and Drug Administration.

Under the provisions of § 58.217, if a sponsor for any reason terminates or suspends a testing facility from further participation in a nonclinical laboratory study and if the test article in that study is the subject of any application to FDA for a research or marketing permit, then the sponsor must notify FDA, in writing and within 15 working days, of the termination or suspension. The notice to FDA must also include the reason for the termination or suspension.

§ 58.219: Reinstatement of a Disqualified Testing Facility

> A testing facility that has been disqualified may be reinstated as an acceptable source of nonclinical laboratory studies to be submitted to the Food and Drug Administration if the Commissioner determines, upon an evaluation of the submission of the testing facility, that the facility can adequately assure that it will conduct future nonclinical laboratory studies in compliance with the good laboratory practice regulations set forth in this part and, if any studies are currently being conducted, that the quality and integrity of such studies have not been seriously compromised. A disqualified testing facility that wishes to be so reinstated shall present in writing to the Commissioner reasons why it believes it should be reinstated and a detailed description of the corrective actions it has taken or intends to take to assure that the acts or omissions which led to its disqualification will not recur. The Commissioner may condition reinstatement upon the testing facility being found in compliance with the good laboratory practice regulations upon an inspection. If a testing facility is reinstated, the Commissioner shall so notify the testing facility and all organizations and persons who were notified, under § 58.213 of the disqualification of the testing facility. A determination that a testing facility has been reinstated is disclosable to the public under Part 20.

CONFORMING AMENDMENTS

At the time of *Federal Register* publication of final GLP regulations, FDA also made amendments to a multitude of other sections of 21 CFR. These so-called conforming amendments all require that a statement be included with respect to each nonclinical laboratory study submitted to FDA in support of an application for a research or marketing permit. The conforming amendment statement can be in either of two forms.

If the study was conducted in full compliance with GLP requirements, the conforming amendments statement will so indicate. If not, then the conforming amendments statement must contain a brief statement of the reason for the noncompliance.

The FDA has required a conforming amendments statement for all nonclinical laboratory studies submitted to FDA after June 20, 1979, the effective date of the GLP regulations. A conforming amendments statement was thus required for studies completed prior to June 20, 1979, if the results of the studies were submitted to FDA after that date.

When several nonclinical laboratory studies are contained in a single submission to FDA, a single conforming amendments statement may be included with the submission, or the sponsor may elect to prepare individual statements for each study.

Preparation of the conforming amendments statement is the responsibility of the sponsor of the study even if the study was conducted by someone other than the sponsor. This is consistent with FDA's view that ultimate responsibility for a study rests with the sponsor. In the case of contracted studies, the sponsor should ask the contractor to supply the information necessary to enable the sponsor to prepare a proper conforming amendments statement. The FDA has not specified who should sign the conforming amendments statement. Generally it will be the same individual who signs the official application for a research or marketing permit. If a statement is included with the report of each study submitted to the FDA, however, the statement may be signed by the study director, laboratory management, QA personnel, or a combination of those individuals.

The FDA has indicated that the conforming amendments statement can be brief for studies, such as preliminary exploratory studies and studies conducted prior to the effective date of the GLP regulations, which are exempt from GLP requirements. In such cases the statement need only indicate the GLP-exempt status of the studies.

Good Laboratory Practice deviations that were of a continuing nature throughout the course of a study will require a conforming amendments statement of the reason for the noncompliance. One-time deviations from GLP requirements should be documented in study records and should be described in the final report but will not require a conforming amendments statement of the reason for the noncompliance.

Care should be taken in the preparation of the conforming amendments statements. While failure to comply with GLPs is only subject to administrative sanctions (e.g., disallowance of a study or disqualification of a testing facility), knowingly submitting a false statement to the FDA is a criminal offense punishable by fine and/or imprisonment.

REFERENCES

1. *Federal Register* 41: 51206, 1976.
2. *Federal Register* 43: 59986, 1978; corrected Federal Register. 44:17657, 1979.
3. *Federal Register* 45: 2486, 1980.
4. *Federal Register* 52: 33768, 1987; corrected Federal Register. 52:36863, 1987.
5. *Federal Register* 54: 9038, 1989.
6. *Federal Register* 54: 15923, 1989.
7. *Federal Register* 56: 32087, 1991.
8. 21 CFR Part 58, Good Laboratory Practice for Nonclinical Laboratory Studies.
9. 21 USC § 371(a).
10. 21 CFR Part 211, Current Good Manufacturing Practice for Finished Pharmaceuticals.
11. Food and Drug Administration. Good Laboratory Practice Regulations Management Briefings-Post Conference Report.

12. Food and Drug Administration, Dockets Management Branch. [HFA-305]. (Available under Freedom of Information Act).

13. SH Willig, MS Tuckerman, WS Hitchings IV. *Good Manufacturing Practices for Pharmaceuticals: A Plan for Total Quality Control*. 2nd ed., revised and expanded. New York: Marcel Dekker, 1982, p. 201.

14. SH Willig, JR Stoker. *Good Manufacturing Practices for Pharmaceuticals: A Plan for Total Quality Control*, 3rd ed. New York: Marcel Dekker, 1991, p. 187.

15. 9 CFR Subchapter A—Animal Welfare.

16. GO Allen, AF Hirsch, H Leidy. Application of military STD 105D sampling plans to research report audits to meet GLP requirements. *Drug Info J* 14:65, 1980.

17. RM Siconolfi. Statistical approaches to auditing. In: *Managing Conduct and Data Quality of Toxicology Studies*. Princeton Scientific, 1986, p. 223.

18. U.S. Department of Defense, MIL-C-45662A, Feb. 9, 1962.

19. SZ Diamond. *Preparing Administrative Manual*. AMACOM, 1981.

20. P Lepore. Presentation at PRIM&R's *Animal Care Committee and Enforced Self-Regulation: Making It Work at Your Institution* conference, in Boston, March 19–20, 1992.

21. U.S. Department of Health and Human Services. *Guide for the Care and Use of Laboratory Animals*. NIH publication no. 85–23, 1985.

3

Applying 21 CFR Part 11 to the Laboratory Environment

SANDY WEINBERG
Muhlenberg College, Allentown,
Pennsylvania, U.S.A.

BACKGROUND

Food and Drug Administration (FDA) regulation 21 CFR Part 11 does not apply specifically to laboratory environments, and does not explicitly modify the good laboratory practices (GLPs). It is difficult to conceive, however, of a laboratory not fundamentally affected by Part 11, and it is all but impossible for a laboratory to interpret and comply with the GLPs without considering Part 11 as a critical context for interpretation. In short, Part 11 has defined the obligations of management in any environment—laboratory, research, production, or clinical testing—in which automated systems are a critical part of the mix. No doubt there is somewhere a laboratory that does not rely on a computer-controlled testing component, does not record information into a database, does not use a statistical analysis software package, and does not rely upon any automation. Increasingly, however, such laboratories are fading into extinction. In all modern facilities, Part 11 becomes a major part of the regulatory mix.

The Part 11 regulations began as a response to the need to accept electronic signatures on computer documents. As hospital and physician records were automated and as electronic identifiers were increasingly accepted in the financial and legal worlds, industry began to agitate for guidelines that would permit electronic approvals of clinical records, standard operating procedures (SOPs), and other documents.

As an FDA–industry task group began to grope with the signature issue, other related guidelines emerged. As finally formulated and disseminated in the late 1990s Part 11 deals with the full range of automation issues, including validation, record retention and storage, auditing trails and change control, access control, equipment configuration, data control, and electronic signatures. Some of the inclusions are simply a formalization and codification of less formal guidelines, as in the validation requirements; other sections represent original requirements, as with the signature, audit trail, and record retention issues. Also, as with any FDA guidelines, it is reasonable to expect further evolution through field interpretations, technological innovations, and transitional solutions.

The Part 11 requirements are fully compatible with the GLPs (perhaps with the slight mental gymnastic of reinterpreting a requirement for a "signature" as an "individual specific approval identification"). The prudent laboratory manager will utilize the Part 11 guideline as an interpretation and overlay of the GLPs. The following checklist and commentary will provide a tool for understanding and utilizing 21 CFR Part 11, and for applying that requirement to a laboratory environment.

CHECKLIST: 21 CFR PART 11

Electronic Records–Electronic Signatures

System is used in support of drug-related research, laboratory analysis, clinical research, manufacturing, production, and/or tracking. Any computerized laboratory equipment, Laboratory Information Management System (LIMS), database system, or other automated system utilized in a laboratory environment is now subject to both GLP requirements and to the Part 11 overlay guidelines. Note that equipment that does not store, interpret, analyze, categorize, or record approval may be exempt from Part 11 issues: it is possible for a laboratory to use even an automated system as an "electronic typewriter," retaining and manually approving paper copies of all records. With these rare (and somewhat antiquated) exceptions, however, Part 11 should be considered an integrated addition to the GLPs.

Record and/or signature system has been subjected to an appropriate and thorough system validation audit. In effect the system validation requirement has been extended to all systems, with the record and signature systems explicitly designated. Some guidelines for validation procedures are included in Part 11. Specifically, a facility is required to establish a specific validation protocol or procedure for its use. That procedure must include: documentation of system development or selection working from a pre-established set of guidelines; both boundary and norm case testing; and utilization of a trace matrix or equivalent tool to tie system requirements to testing. The system must include management controls for change, retention and archiving, and use.

System validation documentation has been collected, including evidence of requirements and design approvals, testing, and implementation. Emphasis is placed upon the sequence, requiring prior approval of system requirements before design of selection; of the design or selection prior to testing and of the testing process prior to implementation. While there is likely to be some flexibility in the first years of Part 11 implementation it should be noted that this requirement implicitly eliminates the possibility of retroactive validation: that is, the assumption is made that all systems in current use have been validated and the possibility of validating systems in prior use has not been incorporated in Part 11.

Records are retained for an appropriate length of time (generally 10 years or two generations over treatment duration) in machine-readable form. This is perhaps the most technologically controversial provision. While it is theoretically possible to retain machine-readable records (ASCII, e.g.—the alphanumeric code—is several decades old), the likelihood that a formatted electronic document will be physically readable several years later is remote. Consider, for example, the 5¼-in. floppy disks of 10 years ago and the fact that new sizes of CDs are currently being introduced.

Retaining records on magnetic tape seems to have some longevity, but the life expectancy of the tape itself is limited. After 5 to 7 years the Mylar substrate becomes brittle, and reading becomes problematic.

Perhaps the safest retention method is pictorial, as with microfiche or microfilm.

Records are retained for an appropriate length of time (generally 10 years or two generations over treatment duration) in human-readable form. Perhaps in recognition of the difficulties described above, this requirement assures redundancy. Paper or alternate forms of human records must also be retained, presumably in an appropriate secure environment.

Records are retained in a heatproof, fireproof, flood-protected environment; are appropriately labeled; and can be restored in a reasonable (generally 72 hr) time. Two subissues beg explanation. Note that the "flood-protected" environment referred to in this interpretation of the requirement is not restricted to places in which a rising creek might threaten retention; the most common cause of flooding is leaking pipes. Certain retention provisions may be necessary, particularly where local fire codes require overhead sprinkler systems. Second, the definition of "reasonable time" for recovery is extremely fluid. A 72-hr guideline has generally been excepted, but there is no uniform industry standard. Laboratories are urged to define recall periods (with sufficient "wiggle room" to cover the unexpected) in their internal SOPs.

Procedures are in place to restrict access to data and records to appropriately authorized persons. Confidential patient records, of course, dictate their own security standards. This form of confidentiality is largely an issue in clinical research, but may impact upon some laboratory environments. Of greater concern here are issues of access related to the potential to modify or delete laboratory records, either through accident or fraudulent intent. A combination of protected audit trails (see below) and "read-only" electronic controls can assure appropriate restriction.

Attention should also be directed to facility security, particularly in laboratories in which controversial animal testing is involved. Visitor identification and restriction in accordance with written policy is appropriate as a protection of both the data that are the focus of Part 11 and of equipment and animal subjects.

Operations checks of the system have been designed in to assure appropriate functioning of hardware and software. The installation qualification (IQ), confirming proper setup of system hardware and software, and the operational qualification (OQ), confirming functionality, have long been included in system validation. While the terms derive from the manufacturing section, these processes are appropriate to and common in laboratory environments as well.

Particular attention is appropriately paid immediately following a system failure or problem. There is concern that improperly close files may result in corruption of existing data, or in misallocation of newly entered data. Most automated laboratory equipment and systems have built-in operational checks and tests to guard against problems. This requirement calls for confirmation of the existence or IQ/OQ features and for the creation of supplemental checks when automated diagnostics are inadequate or absent.

Audit trails (preferably electronic and protected; alternately manual and carefully monitored) have been built into the system to detect and identify data changed, including tracking of time and date of change, change agent; and reason for authorized change. In the "pre-automation" days of manual laboratory systems a bound laboratory notebook provided excellent data security. Authorized changes in data points were recorded with a cross-out and explanatory note. All modifications could be reviewed, and any unauthorized data changes were clearly apparent.

With automated databases it is possible to replace a data point with an alternate value without leaving any apparent record of the change. To avoid possible fraud or error and restore the same level of data security, Part 11 requires the inclusion of an audit trail. An automated audit trail will record any change in data values and provide a field for recording the reason, date, and authorization for the change.

There has been a general field acceptance of a manual audit trail (in effect, a duplicate lab notebook) for established systems that were developed prior to the audit trail requirement. Presumably, however, this acceptance of a manual trail is a short-term alternative.

Electronic signatures are utilized only in systems with dual-level unique identifier authorizations. Early drafts of Part 11 and its antecedent documents have confused this issue. Complex biologic identifiers (DNA readers, retinal scans, fingerprint readers, etc.) are *not* required here. A dual password system, in which a user logs on first to generate a signature and a second time (with a different password) to confirm or affix that signature, is sufficient. There are four general tests of such a system: that all users keep their passwords personal and secret; that passwords are changed with sufficient frequency (perhaps every 3 or 4 months) to maintain that secrecy; that there is a method for controlling override access by system security officers; and that there is a password control system to disable access for ex-employees.

Electronic signatures are utilized only in systems with internal procedures to assure that approved documents have not been modified (without authorization) from a specified date and time. Once an electronic signature has been affixed to a document the content of that document is "frozen," just as the content of a written letter cannot be changed (without evidence of tampering) once it is formally signed. This "locking" is generally accomplished by designating the signed file as read only.

The major complication arises when a global company relies on an electronic signature system. It is possible to electronically sign and close a document in one time zone and to open, read, and perhaps modify that same document a few minutes later in another time zone. Because of time zone differences and the international date line a document may be approved on, for example, January 1 at 11:00, but be based on data not generated until January 2 at noon. These time zone confusions can be controlled by adopting one of three strategies: (1) internally adopting a time/date system based upon an international standard (Greenwich mean time, GMT, most commonly); (2) adopting an internal corporate standard time/date system, perhaps based upon the location of the corporate headquarters (generally referred to as "Zulu time"); or (3) affixing a location indicator (time zone) filed to every dated and timed document.

A *methodology has been implemented to assure the validity of input data. Such methodologies might include dual confirmation of input, the use of check digits, internal norm confirmations, or other techniques.* There has generally been a regulatory acceptance of the validity are entered data: validation concerns itself with maintaining integrity once data is entered. Problems with human blood processing systems, however, have underscored the dangers of a data misentry. This element of Part 11 requires adoption of a strategy to cope with that danger.

In some systems data are entered, then re-entered to confirm. Most computer systems use such a check when users first establish (or change) a password. The user enters the new password, then enters it a second time to confirm accuracy. In the same way a data field can be designed to required dual entry confirmation.

Alternate strategies may include electronic confirmations, such as the use of check digit (your credit cards contain a check digit—usually a factor of two or more other digits—that allows a merchant to quickly confirm the legitimacy of your card number). In other systems, a "norm" violation may signal a possible data error. For example, a system may signal for a data check if a lab animal's weight falls well beyond normal expectations.

Other strategies, idiosyncratic to the situation, can be devices to provide a confirmation of data as they are entered by a human or electronically.

Systems users and administrators have received appropriate regulatory and functional training. Training in job performance and appropriate regulations, including GLP training, is an integral requirement of all laboratory personnel. Part 11 adds to the list of training topics issues of system management and control, including data integrity protection, system use, change control, archiving, and security issues.

System users and administrators have ready and constant access to appropriately comprehensive, clear, applicable, timely, and management-approved SOPs. Standard operating procedures are intended to address two complimentary needs. First, they provide job performance instruction and training reviews to assist all workers in fulfilling their responsibilities. Second, SOPs provide an opportunity for responsible management to periodically review those procedures and to assure that the tasks and methods specified in the SOPs are appropriate, accurate, timely, and complete. Part 11 implies the need for SOPs for validation, security, archiving, training, signature review, data protection, change control, system use, and related topics.

All aspects of the electronic records and electronic signature systems in place have been designed to provide a level of security and control equal to or exceeding the equivalent controls inherent to manual (paper) systems. This "equivalency principle" provides evaluative criteria for all electronic signatures. There is not a mandate to make a system perfect, error-free, or completely "hacker-proof," although these are all appropriate ultimate goals; rather, sufficient controls are required to assure that an electronic system is as secure or more secure than an equivalent manual system.

Consider again the laboratory notebook. Auditing can be accomplished with ease by checking for erasures, cross-outs, overwrites, and other data point changes. Approval is designated with a human signature (which provides a dual level of confirmation: the approver's name and the unique characteristics of the approver's signature graphic). A fireproof file cabinet, properly indexed, permits data archiving and ready retrieval.

The goal of Part 11 is to assure the same level of control in a fully automated laboratory (or other) environment.

COST EFFECTIVE IMPLEMENTATION

Food and Drug Administration regulation 21 CFR Part 11 defines significant and wide-ranging requirements for automated systems used in laboratories (and in other regulated settings). Compliance with the regulation is both mandated and an appropriate representation of good management practices. Unless carefully planned, however, Part 11 compliance can be a cumbersome and expensive—though necessary—process. With appropriate planning strategies, however, laboratory Part 11 compliance can be a smooth, seamless, and cost-effective process.

The modern investigatory, quality control, preclinical, or other laboratory generally requires a highly automated environment. Computer models may be used for preliminary screening; separation technologies are software-driven, and preliminary and preclinical testing involves computer recipes, LIMS networks, and statistical analysis programs. In such an environment compliance with Part 11 and related GLP requirements is not only good science, it is also good business. In a budget-conscious world, however, the business pressures on laboratory management require consideration of cost-effective criteria The well-managed laboratory must comply with the spirit and letter of the regulatory requirements and with extraregulatory concerns related to health and safety, but raising the bar above that which is both required and safety-necessitated adds expense, which ultimately may drive drug and biologic prices beyond affordability for patients in need. Good management, then, calls for cost-effective compliance.

There are four specific strategies for cost-effective compliance with Part 11 in laboratory environments. These strategies include: (1) adopting a multitier validation standard; (2) conducting inexpensive internal compliance audits to identify and focus on key areas of weakness; (3) adopting existing audit strategies to building an internal audit trail; and (4) building compliance into the system selection and adaptation process.

The multitier validation strategy is discussed in depth in the chapter on system validation. In short, all too many companies establish the most stringent of validation standards and apply them uniformly to all systems regardless of actual risk. Adopting a multilevel validation and Part 11 compliance protocol and calling for more rigid controls in high-risk environments is appropriate, accepted, and cost-controlling.

That multitier approach can be effectively married with a strategy of utilization of internal preaudits to identify key deficiencies and areas of real risk. Rather than adopting a policy of implementing controls across the board, the preaudit allows cost-neutral continuation of existing controls where no real problems are found, and allocation of energy and funds to those areas in greatest need of compliance and control.

Third, one of the most expensive changes dictated by Part 11 refers to audit trails. If an electronic audit trail was not designed into an existing LIMS, for example, adding such a trail or replacing the LIMS with a compliant alternative may be a very expensive strategy. Alternately, it is often possible to utilize internal audit procedures, including careful SOPs requiring and assuring the use of a manual audit trail document, to achieve the same level of interim control at a fraction of the expense. Cost is not an excuse for noncompliance, but at the same time compliance does not require choosing the most expensive alternatives.

Finally, clearly reviewing the Part 11 (and other GLP) requirements in preparation for a request for proposal and selection of a new LIMS or other laboratory system can result in cost-efficient building of compliance requirements into an initial system. Again, the audit trail requirement serves as an effective example; building in a secure audit file is a relatively inexpensive addition to the development of a LIMS (or equivalent) file management system. Adding such a feature on after design is a much more expensive process. Maximum cost efficiency can be achieved by using the Part 11 requirements as a part of the selection or design template in purchasing any new automated laboratory components.

With these four strategies compliance with Part 11 can be cost-effectively integrated into the management of any automated laboratory, and while primary consideration must be paid to compliance with requirements and to adding safety controls as necessary beyond those requirements, cost-effectiveness can appropriately be utilized as a third-level criterion in laboratory system selection, design, and management.

SUMMARY

The requirements of 21 CFR Part 11 apply well beyond the laboratory environment, but do stand as supplemental to and interpretative of the GLP in automated laboratory environments. Compliance requires an understanding of the key elements of the Part 11 document, and the formulation of a strategy that assures safety, regulatory acceptance, and cost-effective management.

As the Part 11 requirements are subject to interpretation through field review they will no doubt be modified. Additional changes will result from technological innovation and from the integration of innovative requirements (dual signature approvals, e.g.) into norms of system design, but the intent of Part 11 is critical and permanent.

In a modern laboratory the general guidelines encompassed in good laboratory practices require careful interpretation when applied to a laboratory that relies heavily on automated documentation, data files, and electronic signatures; 21 CFR part 11 provides that interpretation.

4

The Good Automated Laboratory Practices

GERALD J. WHARTENBY, PAUL L. ROBINSON, and SANDY WEINBERG

Muhlenberg College, Allentown, Pennsylvania, U.S.A.

The good laboratory practices (GLP) predate widespread reliance on automated laboratory systems. While the FDA's Part 11 regulations (see Chap. 3) provide additional guidance in part, the U.S. Environmental Protection Agency (EPA) has taken the GLPs a step further. The EPA has authored and released its own good automated laboratory practices (GALPs), providing detailed guidelines for all automated situations. While the GALPs refer specifically to EPA contract labs, they also provide important recommendations for consideration by managers of all regulated laboratories, including FDA GLP labs.

The essential objective behind most data management is control. As such, it is the EPA's ultimate issue in extending its GLP requirements to automated laboratories through the GALPs. The effectiveness of an automated laboratory cannot be assured unless the use and design of the automated systems in that laboratory are consistent with standards intended to assure system control.

The foundation of the GALP standards comprises six principles inherent in the EPA's GLP requirements and its data management policies. These principles define the control issues that caused the development of the GALPs and serve two functions. First, they are the guideposts to understanding the reason behind the GALP requirements and their interpretation. Also, because there are wide variations in the design, technologies, laboratory purposes, and applications of computer systems, the application of these systems is likely to create situations in which appropriate and successful control strategies could evolve that are not anticipated in GALPs. The six principles are thus guidelines for evaluating equivalent options for complying with GALP specifications.

The six principles are

1. *The system must provide a method of assuring the integrity of all entered data.* There is no assumption of system accuracy and performance; rather, a system of demonstrating, presumably through testing, must be in place to provide affirmative evidence of control.
2. *The formulas and decision algorithms employed by the system must be accurate and appropriate.* That demonstration of accuracy must include the operations—the decisions, sortings, and other actions—of the system.
3. *An audit trail must track data entry and modifications to the responsible individual.* As with Part 11, this provision is intended to provide a level of control

equivalent to a paper database. The audit trail should allow analysis of any data changes, including identification of both the reason for the change and the person making the authorization for the change.

4. *A consistent and appropriate change control procedure must be capable of tracking the system operation and application software.* The controls on data are extended to the system of hardware and software that manipulates that data. Here the primary tool for control is a clear and appropriate change control procedure.
5. *Appropriate user procedures must be followed.* Standard operating procedures (SOPs) should provide user guidance and control and should be rigorously followed.
6. *Alternative plans for system failure, disaster recovery, and unauthorized access must be developed.* A problem–response system should provide controls in the event of a significant system failure.

These basic GALP principles have led to the development of a list of specific requirements that provide an appropriate template for effective management and operations of an automated laboratory.

GALP REQUIREMENTS

The purpose of the GALP is to provide a vehicle for demonstrating system control. Control is best exemplified through conscientious adherence to four requirements—documentation, system performance, security, and validation.

Documentation

In general, six types or categories of documents are specified and required for compliance with GALP guidelines. They are

1. Personnel
   a. *Quality assurance (QA) reports* on inspections demonstrate QA oversight.
   b. *Personnel records* help support the competency of various employees assigned to system responsibilities.
2. Equipment
   a. *A hardware description log* records and identifies which hardware is currently in use for a system.
   b. *A record of acceptance testing* demonstrates the *initial* functioning of the hardware.
   c. *Maintenance records* help ensure the *continuing* operational integrity of the hardware.
3. Operations
   a. *A security risk document* identifies likely and possible risks to the security of computer-resident data.
   b. *SOPs* ensure the consistent, controlled use of the system.
4. Facilities
   a. *Written environmental specifications* guard against data loss or corruption from various environmental threats.
5. Software
   a. *A software description* records and identifies which software is currently in use for a system.
   b. *Software life documentation* helps ensure the operational integrity of the software.
6. Operational logs
   a. *Backup and recovery logs* and *drills* help guard against data loss or corruption.
   b. A record of *software acceptance testing* and *software maintenance* or *change control* documents also ensures future software integrity.

General Criteria for SOPs

An SOP must establish guidelines for the specific activities, procedures, and records required to demonstrate and maintain control over the system. Certain criteria must be considered when developing and implementing these procedures.

1. Accessibility
2. Currency
3. Practice
4. Comprehensiveness
5. Credibility

SOPs Specified in GALPs

An automated laboratory requires written SOPs to demonstrate adequate control over automated data collection systems. The minimum SOP topics are

1. Security
    a. System access security and physical security.
    b. Focus: primarily on the computer room and any related workstations.
    c. Physical security: primarily the computer room.
    d. Access security
        1. Access into the computer system.
        2. Modem usage.
2. Raw data
    a. Working definition used within the laboratory.
    b. Restricted access to the raw data archive and the uncorrupted restoration of the data from the archive are prime considerations.
3. Data entry
    a. Identification of person entering data.
4. Data verification
    a. Verification of input data.
    b. Three methods:

1. Double-blind method.
2. Double-key method.
3. Program-edit method.

5. Error codes
   a. Interpretation of codes and corrective action.
6. Data change control
   a. Directed toward minimizing the risk of any unwanted or untested changes taking place within a system.

Safeguards to protect against unauthorized changes and the traceability of authorized changes include

Documentation of how authorized changes have been tested
Proof that the changes do not represent changes that could lead to loss or corruption of data
Cost, scheduling, and impact statements

Must specify the contents of the audit trail and the procedures for printing, reviewing, and archiving the audit log.

The SOPs on software change control and data change control must be written in a coordinated fashion to avoid conflicting requirements.

7. Data archiving
   Must
   a. Be able to store data in a clear, logical, repeatable manner.
   b. Be able to retrieve stored or archived data in a useable, unaltered manner for further processing or analysis.
   c. Specify the detailed methods used to store data, including the frequency of storage, media used, and persons responsible for the storage routine.
   d. Have an indexing system for stored data that provides easy access and record keeping.

e. Specify procedure for retrieving stored data, the authorization process, and procedures for loading it back.

8. Backup and recovery

a. Focus: to ensure the integrity and availability of stored data in the event of a serious breach in security or a systemwide failure.
Specifies
b. Procedures for making and storing backup copies of system data and software.
c. Assigned individual to complete, deliver, and recycle backup copies.
d. Frequency of data backups and the sequence of complete or incremental backups.
e. Types of storage media.
f. Limit to the number of media recycle times and frequency.
g. On-site and off-site backup media storage facilities, again with a delivery and retrieval schedule.
h. Data recovery drill schedules, the dates, people involved, and procedures followed.

9. Hardware maintenance

a. Maximizes the likelihood that hardware will continue to function reliably.
b. If maintenance performed in-house, a responsible person (RP) must be assigned for following schedule and procedure for documenting the performance.
c. If vendor is responsible, an RP must be assigned to document that the maintenance was performed on schedule.

10. Electronic reporting
Specifies:

a. The standards, protocols, and procedures used in data collection and analysis.
b. Format used for reporting data and results.

Additional SOP Requirements

In addition to the SOP topics outlined above, the GALPs specify two other SOP requirements.

1. Each laboratory or study area must have copies of relevant SOPs easily available.
2. All revisions of SOPs and all expired SOPs must be maintained in a historical file, which must also indicate the effective dates of the individual SOPs.

LOGS AND RELATED FORMS

System Backup Log

Used to document regular, incremental, or complete system backups performed in order to safeguard existing data to minimize the future loss of data in the event of a system or application failure

Records the serial or code number of the backup tape, the date of the backup procedure, and the initials of the backup technician

Part of the *backup and recovery SOP*

Routine Software Testing Log

Records the software change identification code or date, the tester's initials, the date and scope of testing, and a pass/fail judgment.

Part of the *software change SOP* or the *inclusive change control SOP*

Software Change Control Log

Used to record all changes made to the system software.

Records the work request code number, the change request date, the system experiencing the change, and a description of the change. It also records testing status, start and close dates, and the programmer's initials.

Part of the *software change control SOP*.

User Problem Log

Used to record user-reported problems with the system or related software
Records the date and problem description, the repair description, and the initials of the repair technician or tester
Part of the *problem report SOP*

System Maintenance Log

Used to record the preventative maintenance completed on particular hardware
Records the date of maintenance, the type of maintenance performed, and the initials of the maintenance person
Part of the *hardware maintenance SOP*

Training Log

This is used to document all user training, including

Orientation training for new users of existing systems
Orientation training for individuals or groups of users of new systems or new versions of existing systems, ongoing training for experienced users
Record of the names and departments of trained users, the date of completed training, the initials of the employee's supervisor, the date of testing or skill review, and the testing supervisor's initials
Part of the *training SOP*

System Operator's Log

Used to record all activities related to the operation of a system
Records the date of activity, a description of the procedure or function performed, and the initials of the operator
Is part of the *physical security SOP*

Security Log

Used to track and identify visitors, consultants, contractors, and other nonemployees who are currently on the premises
Records the date and name of the visitor, the visitor's company, and the times checked in and out
Is part of the *physical security SOP*

Password Control Log

Used to track which users have access to the various clearance levels within the system and to monitor the passwords of all authorized visitors
Records the employee's name and password, the date that password/security training was given, the security level or clearance associated with the employee, the date any entries or changes were made to the log, and the security supervisor's initials
Is part of the *access security SOP*

Data Change Log

Used to record all changes made to data resident in the system
Records the date and time of a change, the system or file involved, the data values before and after the change, the reason for the data change, the initials of the users making the change, and any required approval signatures
Is part of the *data change control SOP*

TRAINING DOCUMENTATION

The GALPs require that a current summary of personnel training, experience, and job description be available for all laboratory personnel involved in the design or operation of an automated system.

The comprehensive and complete training of all personnel interfacing with the automated data collection system must be delineated in a laboratory policy.

A comprehensive employee training program must be established. Documentation must be available that identifies not only the quantity of training each laboratory employee receives, but also the quality of that training.

Training programs must fully document all phases of normal system function as they pertain to the particular user's responsibilities so that each user clearly understands the functions within their responsibility. All training procedures must undergo review at least yearly, as well as whenever new or upgraded equipment or methodologies are installed.

Complete, accurate, appropriate, and available documentation is a necessity for automated laboratory operations.

Personnel

Backgrounds, including education, training, and experience, should be documented and available to laboratory management. Pertinent knowledge of and experience with systems design and operations should be indicated. The important issue is to provide sufficient evidence of training and experience that indicates knowledge suited to job requirements. In light of the need for auditors to verify the qualifications of laboratory personnel, laboratories may consider a separate education and training file for each employee.

Laboratory Management

It is important for laboratory management to

- Develop an organizational plan to document and define lines of communication and reporting within the laboratory structure
- Develop a work plan for any particular study

Laboratory management is responsible to assure that deviations from the GALP standards are reported and that corrective actions are taken and documented.

Responsible Person

The RP must ensure that system documentation in general is comprehensive, current, and readily available to users. In terms of the RP's responsibility for assuring adequate acceptance procedures for software and hardware changes, documentation of acceptance testing can be a part of the approval process preceding the integration of new or changed software into laboratory production. Test data, with anticipated and actual results, should be permanently filed.

Documentation of procedures assuring that data are accurately recorded to preserve data integrity should include audit trail reports indicating all data entered, changed, or deleted. These reports should be reviewed thoroughly by the appropriate personnel.

The laboratory should maintain a written problem-solving procedure, and problems with the automated system that could affect data quality or integrity should be entered on forms or a log following that procedure.

To assure that all applicable GALPs are being followed, the RP should ensure that copies of GALPs are easily accessible, usually in a designated area, to laboratory personnel.

Quality Assurance Unit

A major function of the *QA unit* is to provide proof that the laboratory's automated data collection system(s) operate in an accurate and correct manner, consistent with the recommended function.

The QA unit must

Have a complete and current set of SOPs available and accessible at all times

Have access to the most current and version-specific set

of operations technical manuals or other documentation

Sign off all documentation of inspections

Maintain all records and documentation pertaining to its activities, methodologies, and investigations, including results

Facilities and Equipment

The GALP standards require that a written description of the system's hardware be maintained. Overall descriptions of the purpose and use of the system and specific listing of hardware and software involved in data handling are required.

The systems manufacturer's site preparation manual should be available and the specifications within it must be followed.

Formal written acceptance test criteria should be developed and reviewed before systems are used in production mode.

Specific responsibilities for testing, inspecting, cleaning, and maintaining equipment must be assigned in writing and should distinguish between various hardware devices in the laboratory site.

For each type of hardware device

1. Appropriate testing should be conducted.
2. Written procedures must be followed.
3. A log must be maintained with
   a. Names of persons who conducted tests.
   b. Dates tests were conducted.
   c. indications of test results.
   d. Documentation of any deviations from procedures.
   e. Signatures of management and RP who reviewed.
   f. Testing and results of preventive maintenance by outside vendors.
   g. A list of all repairs of malfunctioning or inoperable equipment.

This must be permanently retained for subsequent reference, inspection, or audit.

Security

Laboratories using automated data collection systems must

Provide security for the systems
Institute a procedure of documented authorization
Establish security files
Appoint a security administrator
Use a visitors' log

Software Performance

Methods for determining that software is performing its functions properly must be documented and followed. User surveys and postimplementation reviews of software performance can be required to evaluate whether or not software is performing its functions as documented.

For all new systems to be used in the conduct of an EPA study, laboratories must establish and maintain documentation for all steps of the system's life cycle, in accordance with the *EPA System Design and Development Guidance* (June 1989) and Section 7.9.3 of the GALP standards.

As far as possible, systems existing in a production mode prior to the effective date of the GALP standards, as well as purchased systems, should be documented in the same way as systems developed in accordance with the *EPA System Design and Development Guidance* and Section 7.9.2 of the GALPs. Documentation relevant to certain phases of the system life cycle, such as validation, change control, acceptance testing, and maintenance, should be similar for all systems.

A written system description, providing detailed information on the software's function, must be developed and maintained for each software application in use in the laboratory.

Written documentation of software development standards must be maintained.

All algorithms or formulas used in programs, including user-developed programs and purchased software packages that allow user entry of formulas or algorithms, must be documented and retained for future reference and inspection. The intent is to establish a source for easily locating such algorithms and formulas.

Acceptance testing of software must be conducted and documented.

Written documentation of change control procedures must exist to provide a reference and guidance for management of the ongoing software change and maintenance process. All steps in this process should be explained or clarified, and the procedures should be available to all system users.

The GALP standards require procedures that document the version of software used to create or update data sets. This requirement is normally met by ensuring that the date and time of generation of all data sets is documented and that the software system generating the data set is identifiable.

Files of all versions of software programs must be created and maintained so that the history of each program is evident. Differences between the various versions and the time of their use should be clearly indicated.

All written SOPs or other documentation relating to software should be available in their work areas to system users or persons involved in software development or maintenance.

Data Entry

Written procedures and practices must be in place within the laboratory to verify the accuracy of manually entered and electronically transferred data collected on automated systems. The primary documentation for data entry requirements is an audit trail. Laboratories must ensure that an audit trail exists and is maintained. This audit trail must indicate date and time stamps for each record transmitted and the source instrument for each entry.

When data in the system are changed after each initial entry

An audit trail must exist that indicates the new value entered.
The old value.
A reason for the change.
The date of the change.
The person who entered the change.

Raw Data

The operational definition of raw data for the laboratory, especially as they relate to the automated data collection systems used, must be documented by the laboratory and made known to employees.

Reporting

When a laboratory reports data from analytical instruments electronically to the EPA, those data must be submitted on standard magnetic media—tapes or diskettes—and conform to all requirements of EPA order 2180.2, "Data Standards for Electronic Transmission of Laboratory Measurement Results."

If laboratories electronically report data other than those from analytical instruments, those data must be transmitted in accordance with the recommendations made by the electronic reporting standards workgroup.

Records and Archives

In addition to specific documentation described above, laboratories must retain all schedules, logs, and reports of system backups, system failures, and recoveries or restores.

All raw data, documentation, and records generated in the design and operation of the automated data collection system must be archived in a manner that is orderly and facilitates retrieval. If stored on the system, such data must be backed up at intervals appropriate to the importance of the data and the potential difficulty of reconstructing it, and the backups must be retained.

Adequate storage space must be available for raw data to be retained in hard copy format or on magnetic media. Storage for system-related records, both electronic and hard copy, must be sufficient to allow orderly conduct of laboratory activities, including complying with reporting and records retention requirements. For the system, this pertains to both on- and off-line storage. Physical file space requirements must be properly planned and managed to meet laboratory needs and responsibilities.

Access to all data and documentation archived in accordance with the GALP standards must be limited to personnel with documented authorization.

Raw data and all system-related data or documentation pertaining to laboratory work submitted in support of health or environmental programs must be retained by the laboratories for the period specified in the contract or by EPA statute.

System Performance

Laboratories utilizing automated data collection systems must provide such control of those systems that current and future system performance can be assured and that data integrity can be maintained. Consistent, accurate, and reliable system performance depends on the control of laboratory facilities and equipment requirements, as well as software requirements. Functional testing, a requirement for both equipment and software, and source code review of software are also required to provide control of system performance.

Facilities and Equipment

The system must be provided with the environment it needs to operate correctly. This requirement applies to all environmental factors that might impact data loss, such as proper temperature, freedom from dust and debris, adequate power supply, and grounding.

Storage capacity and response time must meet user needs.

The RP must ensure that a hardware change control procedure, involving formal approvals and testing, is followed before hardware changes are implemented.

Hardware must be maintained, tested, and cleaned on a schedule that will minimize downtime and problems due to data loss or corruption.

Software Requirements

Each software application in use in the laboratory must perform its functions properly. Determination of continued functionality is related to

1. Acceptance testing—This involves responsible users testing new or changed software to determine if it performs correctly and meets their requirements.
2. Backup—Applications.
    a. Software and systems software must be backed up to prevent complete loss due to a system problem.
    b. Procedures for backups and restores must be established, and reasons should be indicated for which backups other than initial ones should be made, such as changes to software.
    c. Personnel responsible for performing these tasks must be properly trained.
3. Change control procedures—This must be controlled (by the RP) to prevent any changes that have not been properly documented, reviewed, authorized, and accepted in writing.
    a. Variances from any instructions relevant to the system must first be authorized before instituted.
4. Code review—The formulas and decision algorithms employed by the automated data collection system must be accurate and appropriate.

a. Those formulas must be inspected and verified.
b. All algorithms or formulas used in programs run at the laboratory—including user-developed programs and purchased software packages that allow user entry of formulas or algorithms—must be documented, retained for reference and inspection, and be easily located.

5. Audit trails—The laboratory must establish an audit trail so that the software version in use at the time each data set was created can be identified.

Security

Security of automated data collection systems is a major factor in maintaining data integrity. It involves the following three major elements.

Data Protection

Laboratories using automated data collection systems must evaluate the need for systems security by determining whether or not their systems contain confidential data to which access must be restricted. If it is determined that access should be restricted, security procedures must be implemented.

Security *must* be instituted on automated data collections systems at laboratories if data integrity is deemed to be an area of exposure and potential hazard.

Security must also be instituted if the systems are used for time-critical functions of laboratory studies or reporting of study results.

Physical security of the system is required when it stores data that must be secured. All necessary and reasonable measures of restricting logical access to the system should be instituted to prevent loss or corruption of the secured data. The laboratory must establish a hierarchy of passwords that limit access, by function, to those properly authorized individuals who need such functions in the performance of their jobs. Security must be structured in a way that allows access to needed functions and restricts access to functions not needed or authorized.

The laboratory must also establish procedures protecting the system against software sabotage in the form of intentionally introduced software bugs that might corrupt or destroy programs. data, or system directories.

Archiving and Disaster Recovery

The laboratory must establish and follow procedures for system backup and recovery.

The laboratory should develop procedures for applying "work arounds" in case of temporary failure or inaccessibility of the system.

All schedules, logs, and reports of system backups, system failures, and recoveries or restores must be retained by the laboratory.

Transmission

See the section on *reporting*.

The EPA order also provides the formats for six different types and gives other important definitions and information that must be noted and followed by all laboratories submitting data electronically.

If laboratories electronically report data other than those from analytical instruments, those data must be transmitted in accordance with the recommendations made by the electronic reporting standards workgroup.

Validation

Laboratories using computer technology must assure that they have adequate controls in their delivery of data to the EPA. Computer system validation is the process by which a computer system is shown to consistently do what it is supposed to do and only what it is supposed to do. In effect, the validation study confirms and documents the areas of control and the specifications contained in the GALP requirements.

SUMMARY

While the GALPs apply only to the EPA laboratories, and specifically only to EPA contract labs, they provide importance guidance for the manager of any automated regulated lab. Coupled with the specifications of 21 CFR Part 11 the GALPs can serve as important interpretive material in applying the content and principles of the GLPs to the realities of the modern automated laboratory.

5

Implementing GLPs in a Non-GLP Analytical Laboratory

STEPHANIE A. OLEXA
Benchmark Analytics, Center Valley, Pennsylvania, U.S.A.

INTRODUCTION

Good laboratory practice (GLP) standards were initially described in the late 1970s as a set of rules to provide stringent regulatory requirements for research testing of products that fall under the guidance of the Environmental Protection Agency (EPA) and the Food and Drug Administration (FDA). In 1989 the rules were codified by EPA as 40 CFR parts 160 and 792 for FIFRA (Federal Insecticide, Fungicide, and Rodenticide Act) and TCSA (Toxic Substances Control Act) and by FDA in 21 CFR part 58. Currently, these GLP standards differ only to the extent necessary to reflect the agencies' different statutory responsibilities. The standards address nonclinical or preclinical studies in laboratories that do chemical, animal, or field studies in support of applications for research or marketing permits. Compliance has been monitored through a program of laboratory inspections and data audits coordinated between EPA and FDA, with FDA carrying out inspections at laboratories that conduct health effects testing and EPA inspecting laboratories that conduct health effects, chemical characterization, and environmental fate studies. Good Laboratory Practice standards have also been established in over 31 countries across the world, making these standards a universal language of quality assurance (QA).

This chapter will focus on the issues that need to be addressed when setting up GLP standards in an analytical laboratory that does primarily non-GLP work. For most analytical laboratories, either captive (in-house) or independent (contract) facilities, it is likely that only a small portion of the analytical work will fall under the requirements of the GLP, so the decision to implement these standards must be reviewed carefully. Successful implementation of the standards requires a major management commitment of time and resources. Underestimating the challenge of becoming fully compliant is asking for trouble.

WHAT ARE THE GLPs?

Quite simply, the GLPs are a set of rules that are designed to ensure that the data generated by a laboratory support the conclusions that are made. In addition to the published rules, the agencies have provided the advisories that give additional information to help labs understand the application of the rules. The advisories are in question and answer form and address interpretations of the rules. If a laboratory claims to be in compliance with the GLPs the management is stating that the lab staff understands and has implemented all of the GLPs without exception.

WHY DO GLPs?

To comply with GLP regulations is to assure that work is being performed by qualified personnel, under appropriate direction, in adequate facilities, with calibrated and maintained equipment, using standard operating procedures, documenting raw data, having in-process work inspected, and providing reports that are reviewed by a QA professional. This very long sentence sums up all of the key elements of the standards and can be used as the mission statement or motto for a group attempting to set up GLPs.

Until recently, the GLP standards were unique in that they provided a set of guidelines for analytical laboratories that were descriptive and practical. Implementing these guidelines gave the laboratory management and the clients confidence that testing was conducted in a manner that would support the associated conclusions. In the past few years, The International Organization for Standardization (ISO) Guide 25 and NELAP (National Environmental Laboratory Accreditation Program) standards have been promulgated. These guidelines mesh with and complement the GLP standards, therefore for an analytical laboratory striving to reach a high level of technical excellence, the GLP guidelines provide an added tool. Is it appropriate and cost-effective, however, to take on another set of rules?

OVERALL IMPLEMENTATION OR ISOLATED GLP-COMPLIANT GROUPS?

When considering implementation of GLP standards, the first decision for management is whether the guidelines should be implemented across the organization or only in a smaller group of technical areas. Although there are advantages and disadvantages of each, both approaches can be implemented successfully.

If the decision is to implement the GLP system across the organization the cost and time commitment is significantly larger. Training, QA audits, and the general increase in the level of required documentation may involve a major overhaul of the organization with significant disruptions in the normal work flow. This approach provides the advantage that all staff and equipment can be focused on GLP projects when needed, however, giving management flexibility in accepting and scheduling projects.

Implementing GLP standards in a single team or group may be less expensive and faster, but may cause issues if problems arise during a GLP study. For example, if a piece of equipment is validated for GLP standards and appropriate records are kept but the instrument breaks down during a study, a nonvalidated backup unit cannot be substituted. Similarly, if some personnel are trained in GLP standards, other untrained staff cannot participate during a time of an increased workload. Management may also run into problems with a two-tiered staff. In a small organization, keeping the efforts separate may be difficult.

The decision to isolate the GLP units or to integrate the entire organization will depend on the projected amount of GLP projects to be done, the diversity of those projects, the similarity of the GLP standards to other certifications held by the facility, the size of the organization, and the economic impact. Management must consider all the factors before proceeding on a path.

IMPLEMENTING GLPs IN A CONTRACT ANALYTICAL LABORATORY

For a contract analytical laboratory, participating in GLP projects may be a significant diversion from the ongoing business but may allow the company to enter a unique market niche. Many contract analytical labs routinely handle high volumes of samples tested with standardized procedures. Most of the tests are single isolated analyses or groups of analyses, each with a standard operating procedure (SOP) that will apply to a wide range of sample types. The lab generally does not draw conclusions from the tests; it simply reports the values. Audits are usually conducted on methods rather than on a project basis. Customers will range in levels of expertise from highly sophisticated to completely nontechnical. It is not unusual for the technical staff in a contract analytical lab to be unfamiliar with the customer, the history of the sample, or the way the data will be used. The customer may not interact with the lab at all, and may simply be the name on the check. Emphasis is on streamlining and standardizing procedures to get rapid turnaround but reliable and defensible results. Good Laboratory Practice projects require a complete change in mindset and organization. Management and staff should be prepared for culture shock.

The four biggest differences between the routine contract work and GLP projects are

1. The focus on the project rather than on a single method. The project may include a single test or many different types of tests, but it includes all of the testing necessary to reach a conclusion on the test material. This will be completely detailed in the protocol and supported by method and system SOPs. The protocol will be reviewed by the QA auditor and signed by the study director and the sponsor before the project begins. The study director is responsible for all aspects of the project and for formulating the scientific conclusions. Management will have to assign a study director who has the technical capability to oversee all the required tests and to make sound scientific judgments. In many contract laboratories, it may be difficult to identify personnel with the training needed to fulfill the role of study director. The participating staff members will have to coordinate their efforts on the whole project and work less in isolation.
2. The extent of the QA participation in the project. The QA Auditor plays a key role throughout the GLP

project. This person reviews raw data, maintains training records of the staff, reviews the final report and conclusions, and must conduct in-process audits during critical points of the project. The sponsor or client may also assign a QA auditor to the project. Audits will include data review and protocol review, but also in-process audits. The technical staff in a contract lab may not be used to this level of oversight and could see it as intimidating. They may find it stressful to have an auditor watch as the experiments are being done. Management will have to smooth the communication between the QA auditors and the technical staff so that QA personnel can achieve their goals and the technical staff accepts the input as valuable and helpful.

3. The close relationship with the sponsor. Generally, the sponsor or client of the GLP project is also very involved. Many sponsors will conduct a complete audit before contracting with an independent lab. They may also have their own QA auditors conduct site visits and audits during the study. Copies of the raw data and final report must be given to the sponsor for review. Management will face many issues because of this closer relationship. Some sponsors tend to micromanage their projects, which may lead to mixed signals to the staff. When sponsors spend time at the laboratory, management must be cautious to maintain the confidentiality of other clients, whose work may also be in process at the time. Having the sponsor or the sponsor's QA representative underfoot, sometimes during the busiest time of a project, can strain even the best professional relationships.
4. The need to formulate a scientific conclusion from the data. Most contract laboratories conduct analyses and report the information to the client without providing interpretations of the data. In a GLP

study, the study director is responsible for formulating the scientific conclusions from the data. The conclusions must be based solely on the data, taking into consideration the reliability of each data point. Management must support the efforts of the study director and protect him or her from undue influence, including from the sponsor.

Why should a contract lab participate in GLP studies?

Meeting GLP standards improves the overall quality of the work produced by the contract lab and makes meeting other accreditation standards easier.

The work is scientifically challenging and offers the technical staff a chance to excel.

The projects are generally larger and may be more profitable than routine analysis.

The client—lab relationship is closer, leading to a more stable work flow.

Successful completion of projects can allow the lab to access customers nationally and internationally.

What are the risks?

Submission of a project under GLP may trigger an audit from EPA or FDA. Audits are time-consuming and stressful. Both EPA and FDA regulations address the effect of noncompliance with the standards. If a lab submits a study that is found to be in noncompliance, it can lead to a rejection of the study, suspension or cancellation of the permit, and a possible criminal and/or civil penalty. If a test substance characterization is found to be erroneous the consequences can be costly.

Building a reputation and client base in GLP studies can take a long time. Maintaining compliance during that slow growth period can be expensive and difficult.

Keeping a trained QA auditor on staff is expensive. If the lab does not have a full range of GLP projects it may also be difficult to keep the QA auditor challenged and productive. A turnover of QA staff can be detrimental to the quality of the projects.

In conclusion, implementing GLPs in a contract laboratory can be successful and rewarding for the company. There must be a strong management and staff commitment to achieve and maintain compliance, however.

IMPLEMENTING GLPs IN A CAPTIVE OR CORPORATE ANALYTICAL LABORATORY

The decision to implement GLPs in a captive or corporate laboratory is not very different from the decision in a contract-independent laboratory, especially if the lab is run as a separate business unit or profit center. In making the decision, however, management must take additional factors into account.

1. A corporate GLP-compliant analytical laboratory may facilitate the characterization and testing of samples used in efficacy studies. The coordination between the analytical and clinical or field studies may shorten the time to market, giving a significant advantage to the company.
2. Since the GLP projects are more technically challenging and can be more integrated into the overall organization, employee job satisfaction can be improved. This may result in retention of employees in the analytical lab, reducing turnover and decreasing recruitment and training costs.
3. Incorporating global GLP standards into the corporate analytical lab may help the company compete internationally.

A corporate or captive laboratory faces many of the same risks as the contract laboratory. Getting and maintaining management commitment may be more difficult, however, especially in larger organizations.

Again, establishing GLPs and successfully conducting projects in corporate analytical laboratories can be done.

GETTING STARTED IN GLPs

The key element of establishing the GLPs in an analytical laboratory is to set up the quality systems to establish accountability and reconstructability Accountability includes defining the responsibilities of the technical staff, the management, the study directors, and the QA auditors to ensure that they understand and accept the consequences of their actions. Reconstructability means that the systems allow the conclusions to be reproduced from the raw data by another investigator at another time or place.

The quality system includes

1. Management commitment
2. Establishment of a QA unit
3. Adequate facilities and equipment
4. Personnel training
5. Documentation of procedures
6. Record retention and storage

The quality system should be documented in a QA program manual that outlines all the quality policies and procedures for the laboratory. This manual should be considered the handbook for all employees. New employees should become familiar with the document so that they can use it to answer questions that may arise in their work. New or retrained employees should sign a form documenting that they have read and understood the manual. This form should be kept in the personnel files.

MANAGEMENT COMMITMENT

The GLP standards state that the responsibilities of the management are

1. To designate a study director before the study is initiated

2. To replace the study director promptly, if necessary
3. To assure that there is a QA unit that is separate from the study director
4. To assure that test, control, and reference substances or mixtures have been appropriately tested for identity, strength, purity, stability, and uniformity, as applicable
5. To assure that personnel, resources, facilities, equipment, materials, and methodologies are available as scheduled
6. To assure that personnel clearly understand their functions
7. To assure that any deviations from the GLP standards reported by the QA unit are communicated to the study director and corrective actions are taken and documented

Basically, this means that management is *responsible for everything* but *does not do everything*. Who is defined as management, however? In a small contract lab, that may be obvious—the president or technical director of the lab. In a *Fortune* 500 company, is management the vice president of research or the corporate chairman? The GLP advisories shed some light on this issue by defining management as the highest-ranking technically competent person.

Another dilemma for the management of the analytical laboratory is the establishment of a QA auditor who is separate from the study director. In smaller laboratories, the president or laboratory director may be the only individual with broad-based technical skills. This person would be defined as management, however, and as such, cannot act as a study director or QA auditor. The GLP advisories are very clear in establishing that even in a very small contract laboratory, the study director and QA auditor must be separate, with each reporting to management.

For a non-GLP facility to implement these standards, management must therefore commit to hire or train staff who can function as study directors and must have a separate, independent, trained QA unit. Management must take the time to fully understand the regulations and take the leadership role in formulating and implementing the QA policies.

ESTABLISHMENT OF A QUALITY ASSURANCE UNIT

The QA unit can be organized in a way that best suits the individual laboratory as long as the requirement for independence can be met. Some laboratories have a single individual responsible for all QA functions. Others form a committee made up of members from several technical areas. Some small companies may use outside sources to supplement their internal staff. However it is arranged, the QA unit and the QA auditor assigned to a GLP study must be independent of the study director and must report directly to management for this function. These individuals may have other responsibilities in the laboratory as long as management can show that the other duties do not interfere with the GLP work. Training of the QA personnel must be clearly documented to prove that each individual understands both the GLPs and the technical aspects of the project.

PROVIDE ADEQUATE FACILITIES AND EQUIPMENT

As mentioned earlier, when establishing GLP standards in a non-GLP lab, the first decision is whether or not to implement the requirements across the entire organization or only in some technical areas. A key element of compliance with GLP is that adequate facilities and equipment are available for the projects, so the laboratory may face a chicken and egg dilemma, particularly when venturing into GLP projects for the first time.

For a laboratory that will conduct GLP and non-GLP work at the same time, the standards require that the work is kept separate so that the integrity of the GLP study is not compromised. For example, separate analytical standards that are not used in non-GLP studies should be available.

Similarly, a GLP study on the fate of a pesticide should not be conducted using the same equipment as routine analysis of pesticides in soils. Stability studies must be set up in a way that routine work in the lab will not contaminant the test materials.

Equipment used in GLP studies must be validated for appropriateness. Each piece of equipment must have SOPs for operation, calibration, and routine maintenance. All routine and nonroutine maintenance must be documented. What is the definition of a piece of equipment? Any item that can have an impact on the results of an analytical procedure. In the typical non-GLP laboratory, records are kept on analytical equipment such as spectrophotometers or gas chromatography units. Under GLP, however, the definition expands to include items such as pipets, thermometers, incubators, refrigerators, and mixing devices, as long as it is possible that the use of the item can affect the outcome of the test. For the non-GLP lab, implementation of this standard will dramatically increase the number of equipment-related SOPs.

PERSONNEL

The GLP standards require that an adequate number of trained personnel are available for the study. Typical analytical laboratories keep minimal records on the training of personnel, assuming that the quality control results in the individual tests speak for the capability of the analyst. Under GLP standards additional records must be kept and be available for audit. This includes at a minimum

1. The resume of the individual, documenting education, prior job history, publications, presentations, patents, attendance at technical courses, and memberships in technical organizations. The resume must be updated during the course of employment to document additional training and changes in responsibilities. This document should be a brief history of

the employee throughout his or her professional career and current to the time of the GLP study.

2. Training records in the organization. The personnel record should include documentation of training, including safety training, training in GLP regulations, and test-specific training. The records must include when the training was completed, and the topics covered, along with the signatures of the trainer and trainee.
3. A thorough job description for each employee and an organization chart showing the relationship of an individual to the rest of the staff must be documented.
4. All personnel records must be archived and available for audit even if the audit takes place after that individual is no longer with the organization.

DOCUMENTATION OF PROCEDURES

The GLP standards specify that the laboratory must have SOPs for particular activities *at a minimum*. Most laboratories find that implementation of GLPs will cause a dramatic increase in the number of SOPs used by the facility. They are an effective and easy way to document a procedure and to ensure that the staff is trained in the correct procedure. Standard Operating Procedures are living documents and require care and maintenance however.

The first SOP that should be written is the SOP for writing SOPs. This SOP should contain the guidance for the content of each SOP, the numbering system for SOPs, and the system for review, revision, and acceptance of SOPs. Take the time to plan the system so it can grow with the lab and not become too cumbersome.

Standard Operating Procedures should be written by the people who will use them. They should document a procedure with sufficient detail that another individual could recreate the procedure, but not so tightly defined that the analyst will frequently deviate from the procedure. For example, the SOP for an analytical method should specify how standards are made but should allow some leeway for the analyst to use the recipe efficiently on a day-to-day basis. The recipe to make 100 ml of a 50-mg/liter standard needs to be modified with a sentence that allows the analyst to make 50 ml or 200 ml of the standard by adjusting all volumes appropriately.

The original signed SOPs must be in a controlled location, but they must be accessible to all analysts. Paper copies or electronic copies can be used to make the SOPs available. It is very important, however, that the laboratory has a system in place to guarantee that only the most recent revision is in use. All older revisions of the SOPs must be archived so the laboratory has a complete history of the procedure.

RECORD RETENTION AND STORAGE

The GLP standards state that all raw data, documentation, records, protocols, specimens, and final reports generated as a result of a study shall be retained. Specimens do not need to be retained after QA verification. Storage conditions must allow for expedient retrieval with an indexing system. Access to the archived records must be limited to authorized personnel.

For the analytical laboratory in the process of implementing GLPs, the standards will require some changes in record retention and storage policies.

1. Specimen and container retention during and after a GLP study is described in the advisories. The laboratory needs to review these rules and develop an SOP for their facility.
2. The length of document retention and the requirement to provide long-term care for the records in case the laboratory closes must be addressed with a plan.

3. The definition of raw data under GLP is more expansive than in non-GLP work. It includes correspondence, notes, phone records, and any document that relates to the interpretation or evaluation of the data. No raw data can be destroyed under any circumstances.

CONCLUSION

The GLP standards can be successfully implemented in a non-GLP analytical laboratory. The key to success is for management and staff to fully understand the ramifications of the decision and to set in motion a plan to achieve compliance. The plan should include a time line, assignment of responsibilities, and self-audits. The outcome will be an improvement in the overall quality of the analytical work produced by the lab and the ability to participate in a specialized analytical market niche.

BIBLIOGRAPHY

Good Laboratory Practice Standards, 40 CFR part 160, 40 CFR part 792, 21 CFR part 58.

International Standards Organization. Guide 25, 30–8402.

Constitiution, bylaws, and standards. National Environmental Laboratory Accreditation Conference, EPA 600/R-99/068.

RELEVANT WEBSITES

EPA FIFRA GLPS: www.ovpr.uga.edu/qua/epaglp_a.html.
EPA TSCA GLPS: www.ovpr.uga.edu/qua/tscatoc.html.
EPA GLP ADVISORIES: www.ovpr.uga.edu/qua/advisor2.html.
EQP GLP FAQ: www.ovpr.uga.edu/qua/qna.html.
EPA GALPS: www.epa.gov/docs/irm_galp.
FDA GLPS: www.ovrp.uga.edu/qua/fdaglp_a.html.

6

Computer Systems Validation

SANDY WEINBERG
Muhlenberg College, Allentown,
Pennsylvania, U.S.A.

INTRODUCTION

The good laboratory practices (GLPs) provide valuable guidance for the organization and operation of a laboratory. Increasingly the real functionality of that laboratory is dependent upon the accuracy and reliability of a series of automated devices that control instrumentation, data management, archiving, interpretation, and reporting. If those automated systems fail to properly analyze, receive, store, interpret, summarize, or organize data, the integrity of the laboratory can be significantly compromised.

Unfortunately, a combination of poor quality control in the computer software industry, generally inadequate user controls, and the very complexity of the systems themselves have combined to erode confidence in the accuracy and reliability of computer systems. Horror stories proliferate, legitimate regulatory and managerial concerns are common, and the reputation of computerized systems is such that a presumption of confidence is no longer a norm. As in so many regulated areas of laboratory practices, skepticism prevails until support evidence is provided; proof of system control is now required.

That supporting proof of control is termed *validation*. In this context, validation is the demonstration and proof of control of automated laboratory systems, including computerized instrumentation, laboratory information management systems (LIMS), data management systems, and sample control systems. Specific guidelines for the validation of laboratories have not been issued by the U.S. Food and Drug Administration (FDA), though an industry and agency consensus has provided a common understanding of the kinds of supporting evidence required. The U.S. Environmental Protection Agency (EPA) has codified that consensus in a draft guidance document titled "Good Automated Laboratory Practices" (GALPs), which serves as an excellent summary document of the current state of validation throughout the FDA- and EPA-regulated industries.

While the FDA has never endorsed the GALPs (largely for administrative reasons), and while the GALPs do not have the force of EPA regulation, they do provide valuable interpretive guidance and have been widely used by both investigators and field managers. The need for validation of good laboratory practice (GLP) systems has been clearly established, and the GALPs represent a practical, operational, functional definition of that validation proof. For a system to be compliant with specified GALP guidelines, a wide range of controls must be present. The GALPs summarize those tests and controls, with sufficient room for interpretation to meet the varying exigencies of wide-ranging laboratory designs, purposes, and applications. For a system to meet the GALP validation requirements, however, those controls must not only be present; they must be proven. The GALPs not only define appropriate procedures for validation, but also provide criteria for establishing the proof of those validation controls.

The skepticism underlying a demand for proof is not alien to either the scientist or the regulatory professional, yet somehow often emerges as a personal affront when representatives from the two camps interact. Perhaps this resentment emerges from history; the scientist has seen regulatory demands grow beyond reasonable levels, while the regulator has seen behind too many hollow facades claiming to be solid evidence.

In the computer automation field that skepticism may graduate into full-scale cynicism. Technical complexities may exceed the expertise of both scientists and regulators, who have grown increasing uncomfortable with the jargon-filled nonexplanations of the computer professionals. Those computer professionals contribute to the atmosphere, too, with their resentments; their world has never previously had to surrender the shroud of authority for the ego-reducing discipline of double-check and confirmation. Finally, experience has created the need for supporting evidence; too many systems have failed in the past despite all the best promises of control and safeguard.

The result of this combination of history, reality, and attitude is a general regulatory dismissal of any presumption of system control. The "default situation," the unproven norm expectation, is that a system is not adequately controlled. Until firm evidence of that control is provided, an automated laboratory is considered to be without appropriate controls, and both the management and the data of that laboratory are suspect. The GALPs define the controls that are appropriate, and the validation portion of those GALPs define the proof that is necessary to establish compliance.

THE NATURE OF PROOF

Of the classic Aristotelian tripart definition of proof, only two techniques are relevant here. *Logos*, the logical component exemplified in laboratory systems by actual code and function tests, provides important confirmation of compliance. That logos can be verified, tested, and examined. It is the "hard" evidence on which a regulator, or manager, can rely. Included in this category would be the actual logs, test records, and original documents, and similar concrete findings.

Similarly, *ethos*, the testimonial dependent upon the expertise and credibility of the witness, is critical. Evidence supplied by an impartial and credentialed observer may establish compliance with control standard operating procedures (SOPs), accuracy of documentary evidence, and suitability of code design. Whereas the accuracy of logos transcends its interpretation, however, ethos proof must be evaluated on the basis of its source. "Who said so?," "How does he or she know?," and "Why should he or she be trusted?" become the key questions. It is upon the importance of ethos that the important issues of independent, "quality assurance" (QA), and confirmatory investigation lies. Most ethos testimony takes the form of reports, observational records, and certifications.

Pathos, however, the passionate belief of faith, does not apply. A programmer may "know" his code is sound; a manager may be confident her workers are well trained; a supervisor may be convinced the system is reliable. These beliefs are critical, and are not to be disparaged; effective control would not be possible without ultimate reliance upon such well-placed and reality tested faith. Pathos is nonevidentiary, however; it cannot be evaluated independently and falls beyond the realm of science or regulation. Validation must rely on proof; confidence may point to the path toward obtaining such evidence, but is not a substitute for it.

While this may seem a self-evident conclusion, the subtlety of pathos is pervasive. How do we know the system is functioning? The self-diagnostics tell us so. How do we know those diagnostics are accurate? Ultimately, we must rely upon faith, but that faith is not acceptable regulatory evidence, regardless of the passion behind it. Effective evidence, though, buttresses that faith with varying levels of confirmatory evidence: the oscilloscope is calibrated; the testing tool is independently tested; the observer passes the test of independence. Without such checks, data generated by systems cannot be consistently trusted in any scientific sense, and an endless spiral of insupportable claims are left devoid of control.

In the earliest days of computer systems highly inflated estimates of the power, potential, and accuracy of systems created a strong pathos of proof. The "computer says so" became the rallying cry and defense of billing agents, government clerks, and bureaucrats the world over. As stories of enormous and humorous computer errors flooded popular culture in later years, however, a "computer error" became as common a punch line as the "check is in the mail"; computer professionals fell from godlike status to a reputation probably far below the reasonable norm of accurate and reliable system function. The result was, and is, an appropriate demand for controls, even as most reviews demonstrate that those controls are preventive rather than corrective of real problems.

In the appropriately skeptical world of interaction between laboratory scientists and the regulators who must rely upon their conclusions, proof of control must flow from the evidence of logos and ethos. In effect a past history of poorly designed, implemented, and controlled systems has destroyed any pathos to which computer professionals may have otherwise been entitled.

VALIDATION EVIDENCE

Exactly what kind of evidence of validation is required? How much evidence is sufficient to establish clear control? These questions can be answered through an examination of two dimensions. Validation evidence falls into six broad *issue* categories, further defined by two cross matrices of *risk* and *application*. Before defining these two cross-matrix dimensions, though, a detailed description of the issue categories will be helpful.

Evidence of Design Control

Evaluation of any automated laboratory system ultimately involves an assessment of the appropriateness of that system to the job for which it was intended. Regardless of elegance and accuracy, the system is useless if it does not meet the parameters of its application. A bar code system may be intended for tracking samples. No matter how well the software functions, that bar code system is worthless if it does not assign unique numbers and hence fails to allow unambiguous tracking. While such a match seems a self-evident requirement, incompletely considered or changing needs often resulted in systems being used in situations inappropriate to their design.

The key to matching design with system is an effective and up-to-date needs analysis. This process of clearly defining and documenting purpose not only serves to assist in the process of selecting or building systems, but also as a post facto template for managerial and regulatory evaluation of a system. Without a clear statement of exactly what a system is intended to accomplish it is impossible to determine whether or not that (non-) goal is met.

Formal needs analysis approaches often use sophisticated survey and data flow analytical tools to produce a detailed request for proposal from vendors or a comparison model for purchase evaluation. Even the least formal needs analysis must provide three kinds of critical information.

First, the outputs or end results of the system must be clearly defined. In many environments both the format and content of that output is critical. For example, a specific EPA water-testing project may require reporting of lead values, and may also require that those values be printed in a specific location block on a specified form. All outputs should be unambiguously defined, generally through modeling the actual reports or screens that will be required.

Second, the sources of those output elements must be specified. Some outputs are user- (or related system-) entered. For example, a LIMS may receive the water lead levels from a chromatography system. Other outputs may be derived from entered data, perhaps through reformatting the reported lead levels. Finally, some data may be system-generated, perhaps through comparing the received lead level to the average of all other samples and making the determination of whether or not to label a given sample outside the norms.

Finally, the dimensions or ranges of all variables (the outputs and their sources) must be specified. If a system is intended to handle 500 samples per day and can only accommodate 200, it is appropriately criticized. If lead levels are required to three decimal points, a system limited to two decimals is inappropriate. The range of variables is an important specification of system user needs.

These three kinds of information, along with other supporting documentation, must be provided as evidence (logos) of the system design. The review that documentation, assuring its appropriateness, thoroughness, and the degree to which it was followed, provides the additional evidentiary support (ethos) for the system validation.

Evidence of Functional Control

When a system is first installed or utilized, it should be subject to detailed and thorough user testing, including use in parallel to previous systems for a specified period of time. Only when the existing system and new system have produced consistently matching results or when some other comparison process has been used should the new system be considered acceptable. Even so-called standardized software should be subject to this rigor of testing, since unique application or configuration parameters may affect the functionality of the system.

Postacceptance periodic retesting is prudent, and retesting after modification, crash, or problem is all but mandatory. Most of these acceptance and confirmatory tests are user-designed and -implemented, however, providing only limited value as confirmatory evidence. While the tests themselves stand as evidence, the review and analysis of those tests and the review of the test designs require user, developer, and vendor independence for the establishment of credibility.

The validation process provides that ethos by reviewing all test protocols and scripts for thoroughness, appropriateness, and applicability; by replicating a sample of tests to confirm functionality; and independently analyzing the results to arrive at conclusions of acceptable significance levels.

The user tests and validation tests fall into two overlapping divisions: within range (normal function) and out-of-range (stress or challenge) tests. The normal tests evaluate system functionality in expected use. The challenge tests examine performance when parameters of variable, range, and dimension are violated. Ideally, norm tests should show results matching to independent confirmatory sources. Challenge tests should show system rejection of inappropriate data and system maintenance of database integrity despite stresses. Because of the potential for data corruption, challenge tests in particular should be performed on nonlive (library or test) systems.

Evidence of Operational Control

If systems are inappropriately used, the results of those systems are questionable at best. Validation review of a system must include an analysis of proper use and an evaluation of the degree to which normal user behavior falls within those proper use norms.

Norms are established through the development of SOPs, technical operating procedures (TOPs), and working guides (such as help screens and manuals). Those procedures are communicated to users through a combination of memos, manuals, training, and support.

The formal SOPs shall be discussed in further detail in the next section (Evidence of Managerial Control), since they represent the high-level policy decisions of laboratory and system managers. The implementation of those policies is generally specified in the TOPs that detail user activities.

Some laboratories may combine SOPs and TOPs in single documents consisting of a policy and the detailed directions for carrying out that policy. Such a documentary combination is acceptable but not recommended, since it requires a lengthy and unnecessarily complex review process for even the most minor modifications. For example, an SOP may call for safe storage of backup system tapes. A TOP may specify the room to be used for that storage and the inventory procedures for maintaining that room. Should the number of tapes necessitate moving to a second or larger storage room, the TOP can be amended efficiently. If the same change is required within an SOP, a much more complex managerial review process may be required.

The documentation of procedures to be followed, including training outlines and manuals, are an important part of the validation evidence. Accompanying that documentation should be an expert review for appropriateness and a confirmatory observation to determine the degree to which those documented procedures reflect the realities of the laboratory.

Evidence of Managerial Control

In small laboratories the lines of control are simple and straightforward; often the manager and lab technician may be the same person. As laboratories grow in size and complexity, however, the potential increases for a communication problem between the manager of that laboratory and the people involved in basic laboratory activities.

In the regulatory world the manager of a laboratory has a unique role; He or she assumes formal responsibility for the activities and results of that lab. That responsibility is predicated upon the assumption of clear and unambiguous two-way communication; the manager has clearly provided instructions to the lab technician, and the technician has provided effective feedback concerning those directions to the manager. These control issues are significant regardless of the degree of automation in the laboratory. If the laboratory is computerized; however, the control becomes more complex, since the computer in effect becomes an intermediary in the chain of communication. The manager programs or causes to be programmed the recipes and databases for the various tests, which in turn provide instruction to the lab technician. Similarly, the technician enters the data into the system, and the computer provides reports and summaries that provide the control feedback to the manager. With the computer in this intermediary position, managerial control of the system becomes a critical issue in controlling the laboratory and assuming regulatory responsibility for activities and results.

Managerial control is established and documented through a series of SOPs. These SOPs are system design, use, and control policy statements. They summarize procedures of system security, disaster recovery, normal use, data archive and backup, error response, documentation, testing, and other important aspects of control.

Each SOP must meet three tests in order to demonstrate control. First, the SOP must be *appropriate*; that is, a review by management must establish responsibility for the procedures specified, presumably with the evidence of a signature (or in the emerging future, an electronic equivalent). Second, the SOP must be *timely*; that is, the review must be dated, generally within the past 12 months, confirming that the procedure is still appropriate to the situation. Most organizations provide for an annual rereview of all SOPs, including those related to system control. Finally, the SOP must be *available*. All pages must be clearly in the hands of all appropriate personnel, and only the appropriate pages should be in distribution. This requirement presumes some sort of clear recall and control mechanism, some paging control, and some method of SOP storage or posting.

Evidence of Data Integrity

Once data have been appropriately and accurately entered in the system, processed, and stored, they are presumably available for later comparison, analysis, or combination. That presumption is based on the confidence that the system does not in any way corrupt or modify the data, however. Validation requires evidence of continued data integrity.

Four areas of potential threat to data integrity need to be addressed, presumably through a combination of tests, policies (SOPs), and design features. First (and of greatest regulatory interest though probably not very high in reality of threat) is the question of data security. News stories of "hacker" and "virus" attacks of systems have created a high awareness of the potential dangers of malicious or unprincipled attempts to enter a database. Effective protection from security threats has become an important focus of data integrity proof. These protections most often take the form of system locks (physical locks, passwords, software keys, etc.), system isolation (controlled modem access, physical site protection, etc.), and violation trails (logs, audit trails, etc.). In balanced and reasonable proportion these security protections can prevent or detect any threat to data integrity.

Interestingly, too much security can have the undesired effect of reducing protection. If controls are too rigid, making normal productivity difficult, workers have a tendency to develop techniques for circumventing security measures. Complex electronic key doors are wedged open. Passwords are recorded on desk calendars. Systems are not turned off when unattended to avoid complex log-in procedures. In developing security controls a balance with appropriate access must be considered.

Second, disaster situations represent real and potential threats to data integrity. Evidence of appropriate preventive action and recovery strategies must be presented, generally in the form of a disaster recovery plan with an annual practice drill. The disaster recovery plan is usually organized around likely problems (flood from broken pipes, fire, electrical failure, etc.) and includes appropriate notifications, substitute activities, and recovery actions. The disaster recovery plan generally interacts with system backup, recovery, and archive SOPs.

Third, problems of data loss in transmission must be addressed, with evidence of prevention and control strategies. These strategies generally relate to the transmission channels, if any, in effect. The use of bisynchronous channels, bit-checking procedures, and checks digits commonly provide evidence of transmission control.

Finally, data threats related to environmental conditions have generated a great deal of publicity (though in reality are probably very minor). Laboratories located on radon spurs or in or adjacent to nuclear facilities need to be concerned about magnetic and other radiation that may corrupt stored data. An inspection and data reconstruction test generally provides sufficient control proof.

Evidence of System Reliability

All of the areas of proof described above provide evidence concerning the current operations of the computer systems in place, but can those same controls be expected to continue to function over time? Certainly a trend of control provides some presumption, and annual SOP review procedures provide a degree of assurance, but the most significant evidence of system reliability lies internal to the software and is documented only through a review of that source code itself.

Future confidence is based upon the organization of the code, the accuracy of the formulae and algorithms incorporated, and the "elegance" or simplicity of the code. These elements are the focus of the code review.

Poorly organized "spaghetti" code, filled with convoluted pathways that jump back and forth within the code stream, make continued support difficult and create an environment in which future changes are likely to cause unanticipated problems. Alternately, well-organized code allows efficient maintenance with appropriate tracing and variable tracking.

Consistent and proper operation of any software system is dependent upon the decision and action formulae or algorithms included in the code. With a poorly designed algorithm, interim problems may not be obvious in testing, but may cause significant difficulties over time. Similarly, improper formulae may work properly with some data sets, but may malfunction with unusual or "outlier" data points. Examination and confirmation of appropriate formulae is a critical part of any source code review.

Finally, many complex software programs are modified or evolved from other programs. The result may be convoluted dead end pathways, nonfunctioning "dead code," and inefficient module looping structures. Examination of code to determine the elegance or simplicity that avoids these nonparsimonious problems provides an important element in the evidence supporting continued reliability.

The proof in support of reliability is a combination of the logos of the actual code (or reviewed subsection samples) and the credible report examining the elements described above. Here particularly the expertise of the examiner, establishing the thoroughness and soundness of judgment concerning efficiency and reliability of the code, is of particular importance.

THE INFLUENCE OF RISK AND APPLICATION CONSIDERATIONS

The six proof areas described above identify the topics for which evidence must be gathered, but what evidentiary weight is required? How much testing is sufficient? How detailed must a review be? How large a sample of code should be analyzed? When is "enough" enough? The responses to these questions evolve from art rather than science; no absolute definitions are available and no inflexible yardsticks exist, but two parameters provide important guidelines that can be used to generate defensible responses for the vast variety of situations to which the concept of validation applies—"risk" and "application."

Risk refers to the danger resulting from a system-related error. In a blood-processing center, for example, the computer may calculate the appropriate disposition label for a bag of donated blood. If algorithms are incorrect or data are scrambled, a dangerous bag of hepatitis or AIDS-positive blood may be incorrectly identified as safe for human use. In such high hazard situations, testing of the computer system must be comprehensive, thorough, and redundant.

At the opposite end of the spectrum, consider a computer system used to track inert material in a warehouse. Errors in the system may inconvenience the production schedule, but have little or no chance of causing real harm. Even a complete misidentification will be quickly corrected in a QA test of final product. For such a system some validation is still necessary, but a higher tolerance could be used (smaller samples, less frequent rechecks, and broader testing parameters).

The fluid nature of this broad range of risks and the nonspecific relationship to the depth or extent of gathering proof argues further for the expert nature of the process. Only the combination of experience and training that qualifies a true expert will allow consistently appropriate decisions with such inconsistent and murky criteria. The alternative, defining all systems in terms of the most rigid requirements, is an expensive and unnecessarily burdensome alternative.

A further honing of proof quantification comes from the concept of "application." Software can be broadly defined as "standard" (widely used, as with an operating system), "customized" [the multiple site development of a shell defined and written from a standard program, such as the development of a statistical analysis system (SAS) in a C-based language], and "unique" (software written specifically for a single user or site; perhaps, using the same example, the specific protocols written in SAS for use in a specific study).

In principle, the experience of other users with a broadly based system can mitigate the responsibility of any single user. In practice, the need to invest effort in more than a cursory testing of software is eliminated in the standardized packages (except in high risk situations!). For customized software, adjustments in the sample size, depth of analysis, and other factors may be appropriate. As in all safety situations, default should be to the high level; that is, if unsure of the reliability or standardized nature, increase the validation effort. High-risk situations will always argue for increased vigilance, regardless of the number of sites sharing an application.

ERROR LOGS AND PROBLEM REPORTING

The ongoing use and enhancement of a particular application system on a given hardware platform and the installation of additional systems will entail problems and/or failures. In the regulated environment, it is not sufficient to observe that "stuff happens" and continuing processing. There is a special requirement for reporting, classifying, responding to, and resolving problems. This can be the operational companion to rigorous designs and coding standards. Even rigorous system development practices, which carefully document and control design changes, can be defeated by inadequate trouble-reporting procedures. End users who have access to coding or report-generation tools can take it upon themselves to modify and/or enhance what they perceive as an inadequate system.

The discipline of the regulated laboratory requires the equivalent of a notebook or log, physical or electronic, that will record problems. The recording by itself, however, is not sufficient evidence of control. The tracking and resolution of these problems both demonstrate that active measures are being taken to control the system. These entries, linked to activities required to enhance or update the system, provide evidence that the required activities are actually being performed. In addition, they provide an outside auditor with another frame of reference for seeking and reviewing evidence of control.

THE VALIDATION REPORT

The six areas of proof previously discussed also provide a comprehensive package of evidence in support of the GALP. Each area is supported with specific documentary evidence, such as test results, SOPs, manuals, and code, and with testimonial evidence in the form of evaluations, interpretations, and summary reports.

Since the report is in itself a "snapshot" of compliance at a given period of time, it should be updated periodically. A complete revalidation is not necessary, but many sites find that an annual review of the validation report is helpful. Occasional specific events, such as upgrades of programs or replacement of hardware, may trigger partial or complete retesting. Finally, complex systems tend to evolve, so a review to confirm that version control procedures are appropriately followed is recommended on a regular (at least annual) basis.

The report should also establish the credentials of the validating team.

CREDENTIALS

Since the most significant portion of validation evidence rests upon ethos proof, the credentials of the validators are of utmost importance. The credibility of their collective testimony relies upon both their expertise and the objectivity of their conclusions. That expertise is a matter of education and training, experience, and access to appropriate tools and techniques. The objectivity that underlies their credibility, however, is a matter largely of organizational structure.

In any organization a series of reporting relationships define interactions between persons and groups. Those interactions include basic communications, but encompass more complex interactions, including employment and evaluation issues. In the classic QA model a separate and distinct unit outside the normal departmental reporting relationships is used to audit function and activity. The independence of this QA team, free from personal evaluations and budgetary decisions, assures an objectivity of examination. Validation follows the same line of approach. To maximize the credibility of the validation and the value of the testimony provided, validators should be independent of normal lines of authority. Operating as outside consultants or an autonomous QA unit without direct reporting lines to the laboratory or lab management, or through some other mechanism, independence must be assured and proven.

Defining appropriate expertise is even more complex. The FDA informally recognizes the expertise and hence credentials of some individuals, but does not provide any formal certification. That recognition seems to be based upon a combination of experience, academic credentials (in computer systems, regulation, and laboratory management), and general familiarity. (The more often an investigator successfully reviews a laboratory audited by a specific validator the more likely that reviewer will accept the findings of that auditor in the future.) Clearly, then, the credentials of the validator or validators should be established and provided as an important part of the validation report.

Validation establishes the credibility of laboratory data and automated procedures. Without a credible validation review it is certainly possible to follow the GALP guidelines or equivalent industry consensus; validation provides the proof that those guidelines are incorporated in daily and ongoing activities. The GALPs serve two important proposes: they establish the agenda for managing an automated laboratory and they provide a framework for regulatory review of that laboratory's management. Without validation the first purpose can be effectively met; managers can check results, document activities, organize controls, and develop security precautions without any independent check upon their activities. Demonstrating compliance requires validation, however, for it represents the proof that the agenda is followed.

Could regulators conduct their own audits, not depending upon validation by laboratories? In theory that strategy could be successful, but two problems stand in the way. First, resources, including time and expertise, permit only a very cursory spot check on compliance. Those limited resources are much better spent in reviewing comprehensive validation reports than in conducting very limited tests of system performance and compliance.

Perhaps more important, though, is a fundamental philosophical limitation. Is a laboratory manager willing to be so dependent upon a computer system that the only confirmatory check upon automated data is provided by a regulatory inspection? That acceptance would seem to be a real limitation on the kind of control the GALPs, and indeed the GLPs themselves, are designed to encourage. Rather than blindly accepting system-generated results, validation represents prudent checking on system performance.

As a result, validation represents a prudent, cost-effective, and efficient way of assuring regulatory acceptance as well as internal control of automated laboratories and the system upon which they rely.

COST CONTROL

For any company or business unit operating in the pharma arena there are three fundamental obligations.

An ethical obligation—to maintain product and process safety and quality

A legal obligation—to demonstrate that safety and quality to appropriate regulatory authorities
A fiduciary responsibility to stockholders, employees, and customers to meet the first two obligations at the lowest possible cost

Since 1986 numerous spokespersons for the FDA* have urged the system validation of all LIMSs; more than 200 reports of adverse findings (483s) have been issued to laboratories that have not complied. Despite 14 years of FDA efforts to disseminate the requirement to validate, however, the industry is still confused. With the triad of obligations facing a regulated pharma company there is appropriate pressure to provide regular documentation in excess of optimal requirements, compromising the fiduciary responsibility until the minimal bar for compliance is firmly defined.

In short, most pharma companies assure the quality and product safety related to their laboratories, but are overzealous in meeting regulatory requirements to the detriment of cost controls. The responsible segment of the industry is doing more than it needs to do and is spending too much time and too many resources on the validation of LIMS.

STRATEGIES

There are four cost control strategies that if taken together can significantly control the cost of validation LIMS (or other regulated systems) without compromising either product and process quality and safety or regulatory compliance:† the development of a multitiered validation master plan; the regulatory (or expert) review of that plan; the maximization of use

*See for example, the remarks of Mr. Paul L. Figarole, Jr. at the DIA Annual Meeting, Philadelphia, September 1986.

†In fact, these recommendations are drawn from the *Draft Guidelines for Investigators: System Validation* (USFDA, 1998) and private and public remarks of FDA spokesperson Mr. David J. Bergeson, 1998–1999.

of prevalidated, widely used software; and the implementation of an ongoing system maintenance and change control system to reduce the frequency and effort required to periodically revalidate. In one analysis of 112 internal systems in use at a major pharmaceutical manufacturing company, implementation of these four strategies resulted in a savings of $2,100,050 (U.S.)[*] without compromise of either product and process quality and safety or regulatory acceptance.

I. *Multitier validation plan*: The FDA has produced a long list of systems subject to validation requirements,[†] but it is possible and appropriate to limit the amount of investment and effort necessary to compliantly validate each of those systems. Companies are permitted and encouraged to develop a multitier validation plan, using the risk analysis[‡] procedures developed for the regulation of medical devices.[§] Under this procedure a regulated laboratory or facility should

A. Conduct a risk analysis, determining the likelihood and severity of potential consequences of system-related problems. This analysis uses historical data or data from parallel companies to determine what negative consequences are to be expected to occur rarely, moderately, and frequently; and the severity of those consequences (high, medium, or low) in terms of product or process safety and quality.

---

[*]From an in-house IT bid of $2,584,000 to validate all systems completely to an actual cost of $483,950 to validate only the necessary systems according to minimum acceptable criteria.

[†]Basically, every system in the facilities with the exception of financial and human resource systems sometimes calls hazard analysis.

[‡]Sometimes called hazard analysis.

[§]From a regulatory viewpoint a computer system used in research or production of a pharmaceutical or biological product is regulated as a medical device under the Safe Medical Device Act.

B. Conduct a benefits analysis, including quality of life and diagnostic/treatment/cure advantages of the use of the system. Again, both likelihood and significance factors are included in the analysis.

C. Utilize the following guideline to determine the appropriate extent or depth of the system validation effort:
   1. All *high risk* systems, all systems in which *risk* outweighs *benefits*, and unique (custom) software
   2. All remaining systems in which *medium risk* is associated with an equal level of *benefits*, and all software consisting of a prepackaged, widely used core customized or set for unique use
   3. All remaining systems in which *benefits* exceed *risks*; all standardized or prepackaged software

Use of the multitier validation plan allows the effective allocation of resources to appropriate situations in which risks are high or in which there has not been any other testing of the software (custom code). In most environments the implementation of the multitier approach can result in significant savings, concentrating energies and resources on the few critical systems rather than the multitude of insignificant computerized devices and processes.

II. *Regulatory (or expert) review of plan*: Once the multitier validation plan has been developed it is possible to develop a preuse level of confidence by conferring with FDA representatives or experienced industry experts*. While most FDA divisions

*Some experts, involved in the training of FDA field investigators, have become de facto credible reviewers whose opinions and certifications are accepted by FDA field investigators.

will not permit pre-review conferences,[*] the medical device division is generally accepting of such an approach.

In the context of such a meeting it is possible to review individual systems that seem to default into one more rigorous category, but because of special circumstances may actually be appropriately classified in a lower-risk grouping. These meetings are particularly effective in drawing the fine line between "standard with customized features" and "custom systems." Without an expert or regulatory review the multitier approach always defaults to the higher level of validation effort: with such a conference, documented in a finding memo, a finer distinction can be made.

The written plan or SOP on validation should include the regulatory or expert review, identifying the approach as an appeal procedure to resolve issues relating to the three-tier system.

Because of the expense and effort of a comprehensive validation this review may be very cost-effective. The reclassification of a single system from group A to group B may result in saving $50,000 or more without compromising product or process safety, quality, or regulatory acceptance.

III. *Use of prevalidated and standardized packages*: Because software used in the regulated pharmaceutical industries is classified as medical devices it is possible for a vendor of a system to register that package under the Safe Medical Devices Act.[†] The registration process includes submission of

[*]The common response to a request for a meeting: "We are not your consultants."

[†]Effective 1997 such registration is mandatory for systems used in blood processing; other software groups, including LIMS, are likely to be included in the mandatory classification over the next three to five years.

validation evidence, which is then reviewed, presumably accepted, and (under the Freedom of Information Act) made available to users. Other vendors, seeing a marketing advantage, are commissioning third party expert validation studies* and are supplying or selling those reports to their clients or user groups.

While a prevalidation package may not answer all the questions a user has, it can lower the cost of validation in two important ways. First, the use of a widely distributed package can result in a multitier classification into a less rigorous category, and second, the scripts, tools, and findings included in the validation package may provide models and assistance that can save the user a great deal of time and effort.

To a lesser degree a noncustomized package will provide greater confidence than a customized system, if for no other reason than the fact that the problems users find (and force fixes of) are more numerous and of a greater variety. When risks are not an issue, a more widely used system can result in a lower classification.

While this recommendation will have a long-term effect of discouraging innovation, it nonetheless will meet the goal of minimizing validation effort without compromise of quality or regulatory acceptance. All things being equal, select software that is prevalidated or in wide use.

IV. *Effective maintenance and change control*: The validation of a system applies to the code and applications in use at a given time. Almost immediately after the release of a system into normal operations

---

*A validation performed by a vendor is suspect because of a lack of a Quality Assurance norm in the industry: a validation signed by a credible outside expert has much more weight.

(following installation and validation) the system is changed; new reports are developed, enhancements are created, screens are modified, minor errors are corrected, new packages are installed, and delayed features are finally ready. To revalidate after each modification is impractical and prohibitively expensive; even a periodic (perhaps biannual) revalidation can consume resources better used elsewhere.

The solution is an ongoing process of change control and system maintenance, following a preapproved plan and carefully documenting the process. Begin by developing a change SOP that categorizes changes according to their significance; the industry generally uses (or perhaps abuses) a "X, Y, Z" release numbering system.

X: A significant rewrite, major change, or correction of a critical problem; hence, moving from Word 6.0 (or, herein, 6.0.0) to 7.0 connotes a major replacement with new features or structure.

For X changes, your SOP might call for a revalidation in accordance with your master plan.*

Y: An improvement, with the addition of a new feature or features; correction of insignificant problems. Your version 1.0.0 (at time of first use and validation) might change to version 1.1.0 with the addition of a new function or feature.

For Y software changes, your SOP might specify testing the new feature and any modules or sections that interact with that change.

*Watch out for vendors who renumber for marketing purposes. Use a versioning of your own to encompass all the software in a given system. This, your LIMS might include a version of Compex UniLAB4; a copy of MicroSoft Word 7.0; and an Excel Spreadsheet 4.0, but all be labeled your version 1.0.0, unchanged since validation and install.

Z: A cosmetic change, modifying appearance only. Your version 1.1.0 might change to 1.1.1 when a report is remodified to change the order of the columns included or when you have user instructions translated into Spanish for use in a facility in Puerto Rico.

The change SOP might call for testing the change only in the event of a Z modification.

Of course, as in all regulatory issues, if a change seems to fall in either of two categories always default conservatively; a questionable Y change becomes an X, and so forth. You may, as with the master plan, include the consultation with an outside credible expert to resolve any borderline situations.

The use of an effective and category-limiting change and maintenance procedure can lower the frequency of revalidation and can ease the process when revalidation is required. In both situations, significant cost savings can be realized with loss of quality control or regulatory acceptance.

CONCLUSIONS

Regardless of whether the costs of unnecessary and unproductive validation efforts represent a significant portion of a laboratory's annual expense or a small decimal of its budget, managers have a fiduciary responsibility to control costs. If that responsibility is not exercised patients' costs ultimately rise, a portion of needy patients are financially excluded from a beneficial treatment, and investors who might otherwise have funded valuable research and development efforts put their money in an Internet stock instead. Cost control isn't just good business; it's good altruistic policy as well.

In an unregulated environment costs are driven by quality concerns, but in a regulated environment, those costs may exceed quality needs because of regulatory requirements that seemingly demand more control (or more evidence) than is necessary. In a very complex regulatory environment such as the FDA, in which policies must be broad and general to deal with the variety of applicational environments and circumstances, the danger of not understanding the limits of regulatory concern and the chance of exceeding the necessary level of control is great. Uncertain of the fine line of acceptability, the responsible tendency is to overcompensate and do more than is necessary.

In the area of system validation, and of LIMS validation in particular, that has been the trend. Fueled by less than ethical consultants, paranoia concerning the FDA, confusion about the nature of systems, and exaggerated speeches by some FDA representatives, most companies have spent much to learn little and secure less.

The four strategies outlined here—adoption of a differentiating multitier validation master plan, the review and reconciliation of that plan, the preferential treatment of prevalidated and widely used systems, and the inclusion of a rigorous but again differentiating change control process—can minimize the expenditures not necessary to assure quality and to document that quality to regulatory authorities. These four strategies much more narrowly define the minimum height of the bar that legally and ethically must be jumped. The approach allows a company to carefully consider how far above to leap and to carefully control the unnecessary costs of soaring when stepping will do.

APPENDIX: CHECKLIST FOR VALIDATION

Development Process

An acceptable development process includes the elements of

- 1.0 Documentation of process
  - 1.1 Documentation available
  - 1.2 Documentation clear
  - 1.3 Documentation timely

1.3.1 Documentation recorded prior to software development

2.0 Process appropriate

2.1 Process includes all critical steps

2.1.1 Requirements documentation

2.1.2 Design documentation

2.1.3 Coding documentation

2.1.4 Testing documentation

2.1.5 Maintenance documentation

2.2 Process is iterative

3.0 Process is followed

ALL ELEMENTS PRESENT: Accept
ANY ELEMENTS MISSING OR SUBSTANDARD: Reject
DEFAULT: Reject

Requirements

An acceptable requirements step includes the elements of

1.0 Documentation of design

1.1 Documentation available

1.2 Documentation clear

1.3 Documentation timely

1.3.1 Documentation recorded prior to software design

1.3.1.1 Changes subsequent to design should be controlled

1.3.1.1.1 Change communicated as appropriate

1.3.1.1.2 Change approved by management representative

1.3.1.1.3 Change archive for future reference

1.3.1.2 Record is maintained for future reference

2.0 Requirements have been reviewed by appropriate management representative

ALL ELEMENTS PRESENT: Accept
NO CHANGE RECORDED OR MADE, OTHER ELEMENTS ALL PRESENT: Accept
ANY ELEMENTS MISSING (except unmade change record) OR SUBSTANDARD: Reject
DEFAULT: Reject

Design

An acceptable design step includes the elements of

- 1.0 Documentation of design
  - 1.1 Documentation available
  - 1.2 Documentation clear
  - 1.3 Documentation timely
    - 1.3.1 Documentation recorded prior to software coding
      - 1.3.1.1 Changes subsequent to coding should be controlled
        - 1.3.1.1.1 Change communicated as appropriate
        - 1.3.1.1.2 Change approved by management representative
        - 1.3.1.1.3 Change archive for future reference
      - 1.3.1.2 Record is maintained for future reference
- 2.0 Design has been reviewed by appropriate management representative

ALL ELEMENTS PRESENT: Accept
NO CHANGE RECORDED OR MADE, OTHER ELEMENTS ALL PRESENT: Accept
ANY ELEMENTS MISSING (except unmade change record) OR SUBSTANDARD: reject
DEFAULT: Reject

Coder Instructions

Acceptable coder instructions include the elements of

- 1.0 Documentation of coder instructions
  - 1.1 Documentation available
  - 1.2 Documentation clear
  - 1.3 Documentation timely
    - 1.3.1 Documentation recorded prior to software coding
      - 1.3.1.1 Changes subsequent to coding should be controlled
        - 1.3.1.1.1 Change communicated as appropriate
        - 1.3.1.1.2 Change approved by management representative
        - 1.3.1.1.3 Change archive for future reference
      - 1.3.1.2 Record is maintained for future reference
- 2.0 Code instructions have been reviewed by appropriate management representative
- 3.0 Prototyping instructions are included where appropriate
- 4.0 All instructions are sufficiently clear and specific to provide guidance at a level appropriate to the training and experience of the trainers involved

ALL ELEMENTS PRESENT: Accept
NO CHANGE RECORDED OR MADE, OTHER ELEMENTS ALL PRESENT: Accept
ANY ELEMENTS MISSING (except unmade change record or prototyping instructions where appropriate) OR
SUBSTANDARD: Reject
DEFAULT: Reject

Change Control

Acceptable change control procedures should include

- 1.0 Documentation of change control procedure
  - 1.1 Documentation available
  - 1.2 Documentation clear
  - 1.3 Documentation timely
    - 1.3.1 Documentation recorded
      - 1.3.1.1 Documentation complete
        - 1.3.1.1.1 Change communicated as appropriate
        - 1.3.1.1.2 Change approved by management representative
        - 1.3.1.1.3 Change archived for future reference
      - 1.3.1.2 Record is maintained for future reference
- 2.0 Change control procedures have been reviewed and approved by appropriate management representative
- 3.0 A record of change requests and actions is available and archived

ALL ELEMENTS PRESENT: Accept
NO CHANGE RECORDED OR MADE, OTHER ELEMENTS ALL PRESENT: Accept
ANY ELEMENTS MISSING (except unmade change record) OR SUBSTANDARD: Reject
DEFAULT: Reject

Testing Procedure

Acceptable testing procedures should include

- 1.0 Documentation of procedure
  - 1.1 Documentation available
  - 1.2 Documentation clear
  - 1.3 Documentation timely
    - 1.3.1 Documentation recorded

1.3.1.1 Documentation complete
 1.3.1.1.1 Approach specified
 1.3.1.1.2 Sample identified
 1.3.1.1.3 Methodology outlined
 1.3.1.1.4 Acceptance criteria stated
1.3.1.2 Record is maintained for future reference

2.0 Testing procedures have been reviewed and approved by appropriate management representative

ALL ELEMENTS PRESENT: Accept
ANY ELEMENTS MISSING (except unmade change record) OR
SUBSTANDARD: Reject
DEFAULT: Reject

Test Scripts*

Test scripts should include

1.0 Records
 1.1 Key stroke or action taken
 1.2 Expected result
 1.3 Obtained result
 1.3 Decision
2.0 Identification
 2.1 Test date
 2.2 Tester

TEST SCRIPTS AVAILABLE COMPLETE, APPROPRIATE: Accept
TEST SCRIPTS UNAVAILABLE, or NOT COMPLETE, or NOT CONFORMING TO TEST PROCEDURE; or not IDENTIFIED: Reject
DEFAULT: Reject

*If automated test software is utilized, records of results may be in electronic rather than paper archives.

Support and Maintenance

Effective support and maintenance should include

- 1.0 Documentation: Procedure
  - 1.1 Available
  - 1.2 Appropriate
  - 1.3 Clear
  - 1.4 Approved by management
- 2.0 Inclusions
  - 2.1 Release/retest plan
  - 2.2 Notification procedure
  - 2.3 Assistance plan

MAINTENANCE AND SUPPORT PROCEDURE IS DOCUMENTED AND INCLUDES ALL KEY ELEMENTS: Accept
ELEMENTS MISSING OR SUBSTANDARD: Reject
DEFAULT: Reject

Code Review

Effective code review should include

- 1.0 Code available*
  - 1.1 Source code
  - 1.2 Representative sample
  - 1.3 Critical elements (audit trail) included
- 2.0 Inclusions
  - 2.1 Code conforms to coder instructions
  - 2.2 Code is clearly organized in logical pattern
  - 2.3 Code is appropriately documented
    - 2.3.1 Change control
    - 2.3.2 Algorithm

CODE REVIEW IS CONDUCTED AND CONFIRMS ALL KEY ELEMENTS: Accept
ELEMENTS MISSING OR SUBSTANDARD: Reject
CODE NOT AVAILABLE FOR REVIEW*: Reject
DEFAULT: Reject

*A code user may rely upon the results of an independent audit of code as long as that audit followed appropriate steps as outlined.

7

The FDA's GLP Inspection Program

GEORGE W. JAMES
U.S. Food and Drug Administration, Rockville, Maryland, U.S.A.

PURPOSE AND OBJECTIVES OF GLP REGULATIONS

The Federal Food, Drug and Cosmetic Act is enforced by the Food and Drug Administration (FDA) to assure that all regulated products, including food and color additives, animal food additives, human and veterinary drugs, medical devices for human use, biological products and electronic products, are safe and effective for their intended use or uses. The FDA accomplishes this responsibility regarding safety by suggesting the type and extent of testing that is required, by reviewing new product applications to determine whether or not the contemporary scientific standards of safety have been met; and in certain circumstances, by carrying out independent scientific studies to confirm the results submitted by product sponsors. Further to this end, FDA requires that all nonclinical toxicity studies be conducted under conditions that assure that the resultant final report is suitable for informed regulatory decision making. The agency believes that this requirement can be met if the toxicology laboratory is operating in accord with universally accepted principles of good laboratory practices (GLPs).

Each of the five centers of the FDA—the Center for Drug Evaluation and Research, the Center for Biologics Evaluation and Research, the Center for Food Safety and Applied Nutrition, the Center for Devices and Radiological Health, and the Center for Veterinary Medicine—has a special unit that oversees compliance with the GLP regulations (Title 21, *Code of Federal Regulations*, Part 58). The GLP activities of these centers are coordinated in the office of the associate commissioner for regulatory affairs as part of the FDA's bioresearch monitoring program. Both nonclinical and clinical research is included in this program.

The FDA has developed a toxicology laboratory monitoring program to conduct vigorous inspections intended to foster and verify adherence to the principles of the GLPs. The objectives of this program are: to inspect nonclinical laboratories engaging in studies that are intended to support applications for research or marketing permits for regulated products to determine the degree of their compliance with the GLP regulations; to audit ongoing and completed nonclinical toxicity studies to verify their integrity and validity; and to initiate appropriate corrective actions when GLP violations are encountered. The details of the program are contained in the FDA compliance program 7348.808.

TYPES OF GLP INSPECTIONS

There are two types of GLP inspections. The first is the routine inspection, a periodic evaluation of a laboratory's compliance with the GLP regulations. To facilitate scheduling of routine inspections, the agency maintains a list of nonclinical testing laboratories actively engaged in the toxicity testing of regulated products. These laboratories are inspected for GLP compliance at least once every 2 years. The FDA reviews the list for scheduling inspection assignments, and the list is updated when FDA becomes aware of new facilities.

In preparing for a routine inspection, it is necessary to select toxicology studies for audit that are as representative as possible of the laboratory's current operations. This is done either by the assigning center's GLP unit prior to the inspection or by the field investigator at the laboratory site. When made by the field investigator, the selection is drawn from the firm's GLP master schedule sheet.

The GLP master schedule must list all of the studies conducted at the laboratory that are subject to the GLP regulations. This master schedule, indexed by the test article, must describe the test system, the nature of the study, the date the study was initiated, the current status of each study, the identity of the sponsor, and the name of the study director. Using the GLP master schedule sheet, the field investigator may exercise the option to select a study or studies that another of the FDA centers are required to evaluate for scientific content, rather than the studies designated by the center assigning the inspection. For example, if a testing facility to be inspected does not have an ongoing drug study, then a food additive, a veterinary drug, a medical device, or a radiation-emitting product safety study could be selected for audit. In such instances, the GLP staff for the assigning center forwards the information concerning the audited study to the appropriate center's GLP component for review and follow-up action.

The second type of GLP inspection is the directed, or *for-cause*, inspection. The directed investigation is more complicated by its nature than the routine and is less frequently performed in the GLP program. These constitute only about 20% of the GLP investigations completed since the regulations were invoked.

Directed inspections are assigned for one or more of the following reasons:

1. To determine if appropriate actions have been taken by a firm to correct serious GLP deficiencies noted in a routine inspection. This is normally done 6 months after the FDA receives the firm's assertions that corrections have been made.
2. To resolve concerns raised in the preclearance review of final study reports submitted to research or marketing permits, such as an Investigational New Drug (IND) application or a New Drug Application (NDA).
3. To validate critical studies, such as long-term and reproduction toxicity studies, submitted to INDs or NDAs. These studies are selected at each center from master schedules collected in the course of previous GLP inspections or from reviews prepared by the pharmacologist responsible for evaluating applications for research and marketing permits.
4. To verify validations preformed by a third party for the sponsor.
5. To investigate seemingly questionable circumstances brought to the FDA's attention by other sources, such as the news media, other operating firms or laboratories, or disgruntled employees.

OPERATIONAL ASPECTS

Logistically the inspection is a field operation. One of 22 FDA district offices located throughout the 50 states and Puerto Rico will assign the field investigators to perform the inspections. Usually investigators perform routine investigations alone.

Headquarters' personnel, such as representatives of the Office of Regional Operations (ORO) and the Office of Enforcement, pharmacologists of the GLP staff of the assigning center, and on occasion, scientists from the reviewing divisions may be asked by the assigning Center to participate in the GLP investigations.

The ORO acts as a contact for the arrangements involving headquarters' participation in the inspection. The field investigator, designated as the team leader, has the responsibility for the conduct of the inspection and the preparation of the inspection report, known officially as the establishment inspection report (EIR). The lead investigator begins preparation by contacting any headquarters' personnel identified to participate in the assignment in order to make the necessary arrangements for coordinating the inspection.

Another important preliminary to the inspection is the preinspection conference that is usually arranged to include all members of the inspection team as well as any other field and headquarters' specialists judged appropriate by the FDA center assigning the inspection.

NOTIFICATION OF INSPECTION

Prior to 1991, after the inspection team had been formed the next step was for the district office to notify the laboratory of the pending inspection by telephone, about 1 to 2 weeks prior to the inspection. Since 1991, however, laboratories to be inspected are not given advance notice.

AUTHORITY TO INSPECT

The FDA can only enforce inspection of laboratories that perform tests on food, drugs, new animal drugs, or medical device products. Should a laboratory assumed to be doing nonclinical toxicity studies refuse to permit inspection, the laboratory will be advised by the FDA investigator that it is the policy of the agency not to accept studies submitted in support of any research or marketing permit if the agency does not have inspectional information regarding the GLP compliance status of the firm. Even partial refusals, such as refusal to permit access to copying the master schedule sheet and its code sheets, standard operating procedures (SOPs), and other documents pertaining to the inspection, are treated in the same way as a total refusal to permit inspection.

ELEMENTS OF A SURVEILLANCE INSPECTION

The first part of the surveillance inspection covers organization and personnel, which are addressed in Parts 58.29 through 58.35. Investigators must determine whether or not the facility has an adequate number of qualified personnel to perform the types and numbers of nonclinical laboratory studies that it has been (or is), performing. Food and Drug Administration investigators describe in the EIRs the organizational structure and competency of the laboratory. To do this, FDA obtains an organizational chart and the summaries of the training and experience of the managers, study directors, and other appropriate supervisory personnel. If personnel are involved in studies in a location other than that of the inspected facility, the sites and the personnel so involved must be identified. In fact, if there is a need for an inspection of the outside contract facility, this must be specifically noted in the EIR. As part of the organization and personnel evaluation, programs used to increase training and qualifications of personnel through in-house and outside programs must be included in the EIR. As part of this evaluation, FDA must identify, through reviewing the facility personnel SOPs, how the facility recognizes and deals with health problems of the employees, especially those problems that may affect the quality and integrity of studies being performed by that individual.

The quality assurance unit (QAU), the duties of which are described in Part 58.35, presents a special challenge to the FDA investigators. By evaluating QAU activities, the agency is able to assess the mechanisms by which the facility management assures itself that the nonclinical laboratory studies are conducted in a manner that will assure the quality and integrity of the data generated in the laboratory. This is most commonly accomplished by obtaining a list of the QAU personnel and the written procedures for QAU study audits and in-process inspections. The master schedule is also an important tool in the assessment of QAU activities. With it, the investigator can determine whether or not the QAU adequately maintains master schedule sheets and protocols with any subsequent changes or amendments. Food and Drug Administration investigators should always obtain copies of master schedule sheets dating from the last GLP inspection or covering at least the last 2 years. Sometimes the master schedules are voluminous and the investigators may take only representative pages for headquarters' review. Also of interest are the methods by which the QAU schedules and conducts audits. Investigators determine how the QAU retains records and to whom the QAU reports its findings. The records of QAU findings and the records of corrective actions recommended by the QAU and acted upon by management are normally exempt from routine FDA inspection. One exception to the FDA policy of not requiring access to QAU findings and corrective actions recommended and taken is when the agency seeks to obtain these reports during litigation under procedural rules as applicable for otherwise confidential documents.

Parts 58.41 through 58.51 cover the physical facilities of the laboratory. The inspector must determine whether or not the facilities are of adequate size and design for completed or in-process studies. The physical parameters and systems of the facilities as they are used to accommodate the various operations employed in the GLP studies are examined. Investigators also deal explicitly with the environmental control and monitoring procedures for critical areas, especially the rooms used for animal housing, the test article storage areas, and the laboratory areas in which biohazardous material is handled. The procedures and methods for cleaning equipment and areas critical to study conduct as well as the current status of cleanliness are also closely examined. It must be determined that separate areas are maintained in rooms in which two or more functions requiring separation are performed, as well as how that separation is controlled and maintained. The facility inspection must examine the adequacy of pest control procedures, especially in storage and animal housing areas. This is important because residues and improper use of insecticides and pesticides have been known to impact the result of GLP studies.

As would be expected, equipment is also of considerable interest to the FDA investigators. This is covered by Parts 58.61 and 58.63 of the GLP regulations. It must be determined whether or not the facility has sufficient equipment to perform the operations that are specified in the protocols and that such equipment is maintained and operated in a manner that ensures valid results. This is done by examining the general condition, cleanliness, and ease of maintenance of the equipment in the various parts of the laboratory. Also, it must be determined that the equipment is located where it is to be used, and if necessary, located in a controlled environment. For representative pieces of equipment, the investigators check for SOPs, maintenance schedules and logs, and standardization/calibration procedures. It also must be determined if standards for calibration and/or standardization are available. Investigators must be aware of any equipment deficiencies that might result in contamination of test articles, uncontrolled stress to the test system, and/or erroneous test results. Investigators also learn if the same equipment is used to mix test and control articles, and if so, whether the procedures are adequate to prevent cross-contamination.

Food and Drug Administration investigators must give particularly close attention to Parts 58.81 and 58.83, which address the testing facility's SOPs. They must judge whether the studies are being conducted in conformance with these SOPs and in a manner designed to assure the quality and integrity of the data. To accomplish this, they obtain copies of the index and representative samples of all of the laboratory's written SOPs. Furthermore, these SOPs must be available at the locations at which they are to be used. All SOPs and any changes to the SOPs must be appropriately authorized and dated and historical files of SOPs must be maintained. The procedures for familiarizing employees with SOPs must also be reviewed.

Part 58.90 of the regulations deals with animal care. Animal care and housing must be adequate to preclude stress and uncontrolled influences that could alter the response of test systems to test articles. The personnel responsible for receiving and examining animals are evaluated along with the animal care procedures, including any routine treatments, such as vaccination and deworming. Further, FDA examines the criteria used to determine when and for how long animals should be kept in quarantine. Relative to this, GLPs used to separate species and the methods used in handling or isolating diseased animals are examined. At the same time, the method of uniquely identifying newly received animals can be determined.

One of the most important aspects of any nonclinical laboratory study is the preparation and presentation of test and control articles to the test system or test animal. Parts 58.105 to 58.113 of the regulations address this. The FDA reviews the procedures used to ensure that the identity and the dose of test articles administered to the test systems is known and is as specified in the study protocol. In the course of assessing this, the investigators evaluate the methods used in the acquisition, receipt, and storage of test articles. Also, the means used to prevent deterioration and contamination must be evaluated. The identification, homogeneity, potency, and stability of the test articles and the means used to determine these parameters are also closely examined. The methods used to ensure test article integrity and accountability and for retaining and retesting reserve samples of test and control articles must also be evaluated. The aspects of diet mixing that should be observed include: the frequency and methods used to determine uniformity and accuracy of mixing and the stability of test article mixture; the labeling and storage distribution; the disposal of the test article carrier mixture; and the identification and specification of carriers and/or feeds.

Parts 58.120 and 58.130 address the protocol and conduct of the nonclinical laboratory study. The FDA judges whether or not the facility's protocol is generated, approved, changed, or revised in conformance with the GLPs. The overall test system monitoring, specimen labeling, and data collection procedures must be described for the EIR.

The final portion of the GLP surveillance inspection includes examination of records and reports as described under Parts 58.185, 58.190, and 58.195. To accomplish this, FDA assesses the facility's ability to store and retrieve study data, reports, specimens, and so on in a manner that maximizes their integrity and utility. This must include an overview of how the firm maintains materials such as the raw data and the various specimens that are developed in the course of the study. The investigators must become familiar with the facility's archives regarding their location and accessibility. The individuals responsible for the archives must be identified and FDA must learn whether or not the archive is indexed and if the materials and records that have been transferred and stored elsewhere are appropriately identified. Furthermore, the procedures for adding or removing materials from the archives must be examined and individual test systems are selected randomly to determine that all raw data, specimens, and documents have been retained as required.

The examination of records and reports usually concludes the GLP surveillance inspection of a facility, although there may be extenuating circumstances that will prolong the investigation and require closer review of a given area.

STUDY AUDIT

The most important aspect is the audit of completed or ongoing studies. This is particularly true of directed inspections, which essentially is an audit of studies. There are basically two reasons for conducting a study audit during a surveillance inspection. First, there is a need to determine whether or not compliance with the GLP principles by the nonclinical laboratory has resulted in valid studies. Second, it must be determined if a study or studies, either critical or suspect, have indeed been appropriately conducted.

There are 10 prime areas of a nonclinical toxicity study that must be examined.

1. Names, position descriptions, summaries of training, experience, and location of major personnel engaged in the study must be obtained. It is also necessary to examine the workload of selected individuals to determine if they actually had time to accomplish the task specified by the protocol.
2. The QAU for the study must be identified.
3. The QAU schedule, activities, in-process inspections, including the review of the final report and retention of records, must be verified.
4. Significant changes in the facilities other than those currently reported must be closely examined.
5. Any equipment used in the specific study must be examined to determine if it was standardized and calibrated prior to, during, and after use in connection with the study. It must be also determined—if at the time of the study there was equipment malfunction—the impact of the malfunction on the study and the remedial action taken.
6. The SOPs contemporary for the study must be evaluated.
7. The firm's records are examined to substantiate that the protocol requirements were met, and if applicable, the occurrence and types of diseases and clinical observations prior to and during the study must be examined.
8. Any significant changes in test and control article handling from those currently reported are examined.

9. A copy of the protocol is obtained by the team and checked to determine compliance with Part 58.120 of the regulations. It must be determined that protocol changes are properly authorized.
10. A copy of the final study report and copies of interim reports with any amendments must be obtained to determine
    a. Whether or not the final report corresponds with the protocol and describes any subsequent changes in the protocol.
    b. Whether or not the final report accurately reflects the raw data and observations.
    c. Whether or not the final report is appropriately signed and dated and conforms to the requirements of Part 58.185.

SAMPLE COLLECTION

The FDA investigators have the authority to collect samples as described under the compliance program 7348.808. Samples of a test article, the carrier, the control article or test and control article mixtures may be selected and sent to FDA laboratories to determine the identity, strength, potency, purity, composition, or other characteristics that will accurately define the collected sample. In fact, even physical samples such as wet tissues, tissue blocks, and slides may be collected. When the field investigator collects a sample of any chemical substance, he will also collect a copy of the methodology from the sponsor of the testing facility. The copy of the methodology will be sent to the FDA laboratory selected to perform the sample analysis.

PRESENTATION OF THE ESTABLISHMENT INSPECTION REPORT

Before concluding a GLP inspection, FDA officials meet with appropriate laboratory personnel to discuss any observed deviations from GLPs. If there are no departures from the GLP regulations, the facility representatives are so informed during the exit interview and no documentation is given to the firm. If significant deficiencies are found, the laboratory will be presented with a form FDA 483, Inspectional Observations. This form lists the deviations from the GLP regulations as observed by the FDA investigational team during the inspection. When the FDA 483 is issued during the exit interview, the representatives of the laboratory have an opportunity to discuss the statements made therein. The forms may be altered or changed as a result of the exit interview discussions. When issued at the end of the on-site phase of the inspection, the final version of the FDA 483 becomes immediately available under the Freedom of Information Act. As in every inspection performed under the auspices of the act, an EIR reflecting all the findings and discussions is prepared by the lead investigator. The report, unlike the FDA 483, is not available for release to freedom of information requests until all action on the EIR file has been completed.

PREPARATION OF THE ESTABLISHMENT INSPECTION REPORT

The lead investigator is responsible for the preparation of the EIR. Other members of the inspection team may be called upon to participate in its preparation, however, particularly in supplying specialized scientific or technical information. The field investigator and the supervisor at the district office will tentatively classify the completed EIR under one of the following three categories: No Action Indicated (NAI), Voluntary Action Indicated (VAI), or Official Action Indicated (OAI).

THE CLASSIFICATION PROCESS

After report preparation and establishing a proposed classification, the EIR is sent to FDA headquarters with all its attachments and exhibits. The centers' GLP pharmacologists evaluate the EIR and make the final classification of the inspections assigned by that center.

The category NAI signifies "no action indicated." This means that the laboratory is essentially in compliance with the GLP regulations. Ordinarily the inspected facility receives no further correspondence from the agency concerning the inspection, and reinspection is scheduled on a routine basis.

Prior to December 1993, the classifications VAI-1, VAI-2, and VAI-3 were used to characterize the GLP compliance of an inspected facility. The category VAI means "voluntary action indicated." The numerals 1, 2, and 3 formerly indicated degrees of failure to comply with the GLP regulations: VAI-1 meant that the violations were minor and may have been corrected before the inspections was concluded, and VAI-2 indicated that minor procedural deficiencies were found that did not threaten to compromise the validity of any studies done under those circumstances. Those GLP inspections that were formerly classified VAI-1 and 2 are now termed VAI.

Those GLP violations that compromised or potentially compromised the scientific and hence the regulatory merit of a nonclinical toxicity study were classified VAI-3. Those inspections that prior to December 1993 were VAI-3 are now classified as OAI.

The OAI classification has the most serious impact. In such a case, the center of the agency responsible for the appraisal of the test article in question is contacted. A recommendation is made by the GLP staff to the NDA pharmacologists that the study or studies classified OAI should not be used in support of a research or marketing permit, such as an IND or NDA.

In some circumstances, a more severe regulatory and/or administrative sanction is considered necessary by the agency to achieve correction of the violative conditions. For instance, two or more OAI classifications indicating that the laboratory is seriously out of compliance could result in the disqualification of the laboratory (Title 21, CFR, Part 58, subpart K).

Classifications of OAI would be considered when any one or more of the following exists:

1. Quality assurance is poor or nonexistent.
2. The test article and its dosage forms have not been characterized as required by Parts 58.105 and 58.113.
3. A study or studies that must comply with the GLPs have not been listed on the master schedule.
4. Numerous less serious GLP deviations that persist over two or more inspection periods. This suggests that the laboratory is out of control.

In the case of OAI classifications, a directed reinspection will normally be assigned on a schedule determined by the center initiating the investigations.

GLP INSPECTIONS ABROAD

It may rightfully be said that all this information is interesting in terms of laboratories in the United States, but what about GLP inspections abroad? What steps has the agency taken in this direction? When the FDA developed the GLP regulations and its laboratory inspection program in 1976, the planners were preoccupied with domestic laboratories. A survey of investigational drug submissions completed since then found that approximately 42% of the safety studies submitted had been conducted aborad. This convinced the FDA that safety studies conducted in foreign laboratories would have to be addressed.

While the laws of this country require that the safety of foods, chemicals, and drugs be demonstrated by well-controlled studies, the authority of the FDA cannot be exercised beyond the borders of the United States. Concluding that the best means of satisfying the law would be to physically observe the operations and practices of the laboratory, the agency took it a step further by announcing that a refusal to permit such an observation would result in the nonacceptance by the FDA of the uninspected data. This standard applies to all laboratories—foreign and domestic, governmental as well as commercial—that conduct studies intended for submission to the FDA.

Since beginning inspections of foreign laboratories in 1977, the agency has visited about half of the approximately 110 foreign laboratories that have conducted studies that have been submitted to the FDA. The FDA has inspected laboratories in most European countries as well as in Japan and Australia. Because these inspections are relatively expensive, the FDA's focus on international inspections is directed to laboratories that are frequent contributors of critical studies to research or marketing applications. Up to this time, the FDA's foreign GLP inspection teams have found excellent cooperation extended by the foreign laboratories. There have been no refusals to inspect, and the quality of the studies audited is no better or no worse than the quality of similar studies conducted in the United States. Mainly, the GLP problems were inadequate SOPs, discrepancies between raw data and the final report, undocumented protocol changes, and improper correction of recorded data.

As already mentioned, foreign GLP inspections are extremely expensive. To avoid incurring these costs, the FDA has made bilateral agreements with the drug regulatory agencies in several other countries.

Phase I commits the drug regulatory agency of each country to establish a GLP program, provide for joint inspections, and share information and consultation. Once a GLP program has been established, an assessment is then made of the program to establish comparability between the inspection methods used by the foreign regulatory agency and those used by the FDA. Phase I agreements are presently active with Sweden, Japan, and Canada.

The phase II agreement, when reached by participating countries, affords reciprocal recognition of each country's program and provides for mutual acceptance of data and exchange of inspectional findings. At the present time, the FDA has phase II agreements with Switzerland, Italy, France, Germany, and the Netherlands.

THE RESULTS OF GLP INSPECTIONS BY THE CENTER FOR DRUG EVALUATION AND RESEARCH

Having discussed the inspection processes of FDA as far as GLP regulations are concerned, the question may then be asked: What are the recent and historic results of the FDA's toxicology laboratory monitoring program?

For fiscal year 1993, 47% of the 80 GLP inspections classified as involving human drugs resulted in NAI classifications. Of the 80 inspections, 29% (23) were classified VAI-2; 9% (7) were VAI-3 (VAI-1, 2, and 3 classifications were used prior to December 3, 1993); and 8% (6) resulted in an OAI classification. These percentages are based on the GLP inspections assigned only by the Center for Drug Evaluation and Research and do not include a small number of VAI-1 classifications in which violative conditions were corrected by the laboratories prior to completion of the inspections. Since the inception of the regulations on June 20, 1979, to March 31, 1993, the center has reviewed a total of 931 inspection reports. There were 408 inspections of sponsor labs, while the like values for contract, university, foreign, and government laboratories were 435, 57, 25, and 6, respectively. The majority of the EIRs from these inspections (in excess of 80%) were classified by the centers as NAI, VAI-1, or VAI-2. These results reflect favorably on the positive attitude of industry in implementing the GLPs. During the same period, 124 reports were classified VAI-3, and 47 were classified as OAI.

It should be noted that during this period FDA 483s were issued in 498 of the investigations, more than half of the total. It must be kept in mind, however, that the FDA 483 lists only the observed deviations from the GLP regulations; it does not prioritize the seriousness of the deviations. The significance of these observations is determined during the review and classification of the EIR at headquarters. Furthermore, the fact that an FDA 483 was not issued does not imply that the firm was in compliance. In three instances during fiscal year 1993, when an FDA 483 was not issued, the agency sent a letter based on the center's evaluation of the inspection report alone. Based on the EIR, it had been concluded that although no FDA 483 was issued, the findings in the report were important enough to be communicated to the laboratory.

Of the 931 inspections classified, information letters were sent 395 times, notices of adverse findings letters were sent 106 times, and letters stating study rejection were sent 49 times.

LABORATORY GLP COMPLIANCE RATINGS

Some may question how the inspected laboratories rated in terms of compliance to each of the 141 operational provisions of the GLPs. The information accumulated from the Center for Drug Evaluation and Research indicates that 66% of the inspected laboratories were cited for one or more deviations from these provisions. The most significant departures from the GLPs were: (1) final reports did not conform to the raw data; (2) improper correction of the raw data; (3) protocol revisions were implemented without amending the protocols; (4) the absence of required SOPs and the failure to amend SOPs when necessary; and (5) the master schedule sheets and the protocols did not contain the information required by the regulations.

Food and Drug Administration investigators found generally acceptable performance in the archival and record retention areas as well as in the area of the physical facilities associated with animal care and laboratory operations. The lack of findings in these areas is encouraging, since it may be recalled that a major problem that precipitated FDA's concern for the quality and integrity of safety data was in the area of raw data retention.

GLP IMPACT AND NEW DRUG EVALUATION

With all of the foregoing information on the GLP inspection procedures and the statistical evaluation of the completed inspections, one must still ask "How have the GLP inspections impacted new drug evaluation?"

First, the people responsible for FDA's bioresearch monitoring program are encouraged by the results of the GLP inspection seen in terms of industry's growing acceptance of the GLPs as a means of establishing a level of reliability for scientific testing.

Furthermore, we know that the deficiencies found by our inspections in the past year are not as severe as in recent years and the cooperation we are now receiving from laboratories during the investigations is at a higher level.

Finally and most important, the pharmacologists at the agency, particularly those who are keenly aware of the conditions that existed before the GLP regulations came into effect, are in agreement that the GLPs have made the reviewer's tasks much easier, and they, the reviewers, feel more confident of the reliability of the information that comes across their desks.

8

The Future

SANDY WEINBERG
Muhlenberg College, Allentown,
Pennsylvania, U.S.A.

The first two editions of *Good Laboratory Practices* concluded with chapters of predictions about the future of GLPs and laboratories in general. A review of those previous predictions suggests that the crystal ball has proven remarkably reliable. In reality, of course, a crystal ball has not been the instrument of choice. Much more reliability can be obtained by examining the entrails of goats.

While acknowledging tongue in cheek, consider for a moment the entrails strategy. Other oracle methods rely purely on magic. It is difficult to postulate even an indirect scientific explanation of the predictive power of tea leaves, clover, or the crystal ball, but if a goatherd approaches and asks for a prediction of his future, examining the health of one or more randomly selected members of the flock is a sound strategy. If the selected animals are disease-ridden, the goatherd is not likely to experience future prosperity. Conversely, a healthy specific is an effective predictor of a steady source of goat milk, cheese, meat, and other products and the economic boom they produce. The entrails strategy is really a technique of extrapolation—of predicting not by magical means, but by extending the realities of today into the near future.

Using that strategy of extrapolation, without the mess and fuss of sacrificing any animals, there are some very reasonable predictions that can be made about the future of the GLPs.

THE GROWTH OF AUTOMATION

Perhaps the only surprising element in a prediction of increasing laboratory automation is that there is room for any additional growth; many labs seem to be fully automated. Of course there are virtually nonautomated laboratories that are likely to catch up to the state of the art within the next few years, but there is an automation trend that is just beginning to emerge and is likely to represent the most dramatic growth area. Increasingly, laboratories will automate the *regulatory function*.

The first significant innovation of automation of laboratory regulation is found in 21 CFR Part 11. Arguably the most innovative part of that regulation is the requirement for an audit trail as an integral part of every database. Consider the evolution of audit trails. Originally, an outside (Food and Drug Administration; FDA) investigator examined the laboratory notebooks in use to assure the accuracy of recording and the integrity of what has become known as the "database." Effective laboratory managers, motivated by quality assurance (QA) altruism and by the threat of FDA investigation, instituted their own internal audits, which were eventually relegated to an independent QA department.

As databases were automated that review function became increasingly problematic, and increasingly a concern of both outside FDA and internal QA auditors. As a response to the difficulty in tracking a data point change in an electronic file automated audit trails were developed, encouraged, and finally (in Part 11) required.

A subtle shift has taken place, however. Ten years ago a laboratory might have been (appropriately) cited for poor data protection practices. Now a citation might be issued for lack of an audit trail, even with evidence that the data have never been compromised. In effect the regulatory attention has shifted back a step, from verification of data to verification of data protection devices (audit trails). The FDA now requires an automated regulatory tool as a QA monitor in all but the most unautomated laboratory environments.

This shift to requirement of automated regulatory functions is likely to continue, in part because of our increasing reliance on automated processes and in part because well-designed monitoring tools make good sense to both the mission of the FDA and the goals of the industry. It is reasonable to suggest that the next evolution in the GLP laboratory will include automated monitoring equipment performing and recording continuous checks on data entry functions, analysis algorithms, sorting routines, and other data control functions. In time these automated regulatory checks will become as routine—and as mandated—as temperature gauges on refrigeration units.

LAB IN A BOX

The one element of greatest quality and compliance vulnerability is still the human being. In even the best GLP laboratory lab workers can purposely, inadvertently, or accidentally subvert quality controls and operating procedures. Largely motivated by this vulnerability, and secondarily by the steadily rising relative cost of labor, a number of laboratory functions are being fully automated to avoid any human element at all. A field water testing system has been developed, for example, that requires human involvement only in the initial input of the water sample; all of the remaining sample separation, testing, interpretation, and reporting is performed by robotic elements in the suitcase-sized automated laboratory. A similar robotic blood testing laboratory, automating all functions once a test tube of human blood is entered, is designed to perform 12 separate tests, to apply the result to a matrix template, and to print out an analysis and a label classifying and describing the sample. The entire unit is about the size of a laptop computer. Tests at several university hospitals have proven highly promising.

NASA has designed and contracted a water testing, purification, and retest (certification) system that is a self-contained unit about the size of a small automobile. Human involvement is restricted to putting "hose A" into a puddle, polluted stream, saltwater bay, or almost any other liquid source, and to drinking pure water from "hose B." The fully automated laboratory (and processing) system is intended for an eventual Mars base. It is currently used on navy ships and some desert warfare army units.

As the technology improves and quality controls are built in we can expect additional available "laboratory-in-a-box" options. Such systems, perhaps registered with the FDA under the 510(k) Safe Medical Device Act, would shift a good deal of the regulatory efforts from user to manufacturer. Increasingly the GLPs are likely to assist the design engineers of such self-contained robotic lab units as they have the managers of traditional laboratories.

CENTRALIZED LABORATORIES

Even as increasingly sophisticated self-contained GLP compliance lab-in-a-box options become available, another countertrend is likely to result in fewer traditional labs as better data segregation techniques make centralized laboratories an attractive alternative.

In clinical research new regulatory emphasis, increased patient management requirements, and reliable data segregation technologies resulted in the emergence of freestanding research companies that could design, conduct, and statistically analyze drug trials. Similarly, the clinical testing laboratory industry emerged as tests became too sophisticated for physicians' offices, regulations grew too cumbersome, and advanced technologies became available. These same pressures are likely to result in fewer, larger, more automated centralized GLP labs. In some larger companies these labs may be operated as departments within the corporation, much like the Information Technology (IT) department. In other circumstances it is reasonable to expect the emergence of freestanding service companies offering laboratory functions on a time-share basis, guaranteeing regulatory compliance.

Centralized laboratories offer several advantages. First, and perhaps most obviously, centralized operations can afford greater capital investment in state-of-the-art automation systems, top-level employees, and the most sophisticated equipment. While upgrading a Laboratory Information Management System (LIMS) on a regular basis may not be cost-effective for a small QA laboratory, it will be a practical annual expense for a company offering QA lab services to a number of clients.

Second, centralized facilities maximize the efficiency of the regulatory process. While the FDA may send a generally trained investigator to review a production plant on an occasional basis, it can provide regular investigations by top trained specialists for a centralized laboratory company. More frequent and more expert investigations work to the advantage of both the industry (experts have a more sophisticated understanding of appropriate regulatory interpretations) and the general public (closer regulation and review of self-regulation maximizes compliance and safety).

Finally, centralized laboratories are a significant step toward the eventual goal of limited data sharing. Currently there is a potential (and real) possibility that several companies, each with experience with a different representative of the same class of drugs, has discovered an occasional adverse reaction. That reaction may be so infrequent, however, as to appear a statistical anomaly of no importance. Currently the data would be viewed in aggregate only in the COSTART (combined reporting) system. In a centralized lab there is an increased chance of observing a trend at its earliest stages, particularly as laboratory users begin to accept the cost savings of using of overlapping control groups.

Taken together these factors suggest an improved quality level in centralized laboratories. The centralization trend is likely to be a result of financial pressures; the result, however, is likely to be better control

NEW REGULATORY FOCUS: DATA QUALITY

The history of the FDA is a story of continuously "stepping back." Is the product free of contaminants? Then step back and consider the process of production. Is the product well organized and safe? Examine next the hardware and software systems that control that manufacturing process. Every time one level of concern has achieved industrywide predominance it is possible to apply the energies used to achieve that awareness and compliance to yet another previous step. The process is, in effect, continuous quality improvement in action.

In the 1990s the major focus was on system validation, on assuring the quality control (QC) of the hardware and software that operated the manufacturing process, automated the laboratories, and controlled the inventory. As that validation process achieved mainstream acceptance and majority saturation it is appropriate to ask "What next?" Where can we and the regulators most productively next turn our attention to maintain continuous quality improvement?

The clue comes from one of the old truisms of the computer field: GIGO (garbage in, garbage out). Translated, computers can accurately and reliably process, interpret, store, retrieve, and sort data, but the quality of the result is a direct effect of the quality of the data originally entered.

System validation has always focused on the integrity of data once they are entered into a system, but if a research scientist misenters a data point or a robotic system misaligns a test tube, data that meet all validation tests but are erroneous or misassigned can corrupt the end result. In effect, if your finger slipped and you typed in a "3" when you meant a "2" your computer will still add properly, but the answer you receive will be incorrect—GIGO; the ultimate accuracy of laboratory systems is based in part on the accuracy of the data that are entered.

In 21 CFR Part 11, in effect a modification of the GLPs (and other guidelines), the quality of the input data is mentioned, but little emphasis is placed on this important phase of QC. As compliance is achieved in other areas, however, it is reasonable to assume that regulatory attention will increasingly focus upon techniques for double entry, norm range checking, and other data quality checking strategies. Over the next few years, as compliance with Part 11 reaches the current levels of other GLP compliance, the focus is likely to shift back once more to the quality of the actual data and to the data entry process.

OSHA INVOLVEMENT

To date the Occupational Safety and Health Administration (OSHA) has had little involvement with GLP labs other than as a part of normal workplace safety evaluations. Increasingly however, OSHA has expressed concerns about the possible exposure of workers to biological and viral contaminants in laboratory environments. No doubt a good deal of this concern is a result of concerns following the September 11 disaster, and subsequent (so far unrealized) fears of related biological warfare.

Over the next few years this concern is likely to take two important forms. First, while no formal guidelines have been announced, there are reports of informal discussions within OSHA about possible new and more stringent laboratory safety controls. To date OSHA has relied largely on self-policing of laboratories in accordance with GLP principles and with guidelines from industry-support laboratory groups. The American National Standards Institute (ANSI), the Alberta (Canada) Society of Medical Laboratory Technologists, and the Research Defense Laboratory Association, for example, have quasi-voluntary worker safety programs.

Many state and federal clinical laboratory (CLIA) programs have included worker safety criteria, and the Canadian Office of Laboratory Security—strengthened in response to the possibility of bioterrorism—has added a laboratory safety division. Based upon these models it seems a reasonable prediction that OSHA will soon launch the process of coordinating and consolidating worker safety laboratory standards into a new and comprehensive code.

Second, prior to, during, and subsequent to the code development process, it is safe to predict that OSHA investigators visiting facilities that include laboratories will begin to focus attention on worker safety in GLP facilities. Laboratory worker conditions are likely to get priority concern.

While the goals of worker safety are certainly admirable, emphasis on this area by OSHA is likely to cause some disruption in GLP labs. First, any set of overlapping regulations—OSHA and GLP, for example—is bound to cause some confusion and contradiction. Second, OSHA investigators, no matter how well trained and dedicated, are generalists with little in-depth understanding of the intricacies of laboratory operations. Their interpretations, particularly early in the learning curve, are likely to be problematic.

The answer, of course, is increased inclusion within the GLPs of worker safety issues, in effect making OSHA involvement unnecessary. Either before or as a result of problems resulting from increased OSHA involvement, it is safe to predict further GLP emphasis on safety.

KEY ROLE

One final prediction bears repeating from the previous edition of Good Laboratory Practice Regulations. As was true 4 years ago and as is true today, it is safe to predict that over the next few years the regulation of laboratories, governed and guided by the GLPs, will continue to play a key role in assuring public health and safety. In part this key role is a result of the significance of the GLP laboratory in the QA function related to all phases of pharmaceutical operations. In part it is a function of the increased complexity of the GLPs environment itself.

Good Laboratory Practice labs play an important part in the research and discovery (R&D) function, from molecular modeling through screening of promising compounds. Modern R&D is largely a matter of focusing efforts on molecules and compounds of the greatest potential. It is the well-organized laboratory, compliant with GLP principles, that is least likely to reject pursuit of promising drugs and most likely to lead to potential breakthroughs without fear of result distortion through misinterpretation, contamination, or database corruption.

Preclinical and clinical process laboratories serve a critical function in the development and organization of toxicology data, animal testing results, and ultimately of human test results. Again, the GLPs provide guidance and assure both minimization of misinterpretation and control of both the potential risks to human subjects and the suffering of test animals.

In the manufacturing process, the QA and QC laboratories are fundamental in assuring sterility, purity, labeling accuracy, and other aspects of quality. The GLPs provide the template for control to assure the reliability of results and the credibility of the final product.

In all these stages, the increased complexities in the use of biologics and the use of more complicated compounds and more sophisticated testing and manufacturing technologies assure increased reliance upon the laboratory and upon the

GLPs that define effective and efficient laboratory management.

Coupled with the increased importance of GLPs resulting from the increased importance of the role of quality in a complex environment is the changing nature of that very environment. As all aspects of the process rely increasingly on automation two phenomena emerge. First, automation itself adds complexity as databases are manipulated, massaged, graphed, and statistically analyzed. Second, and perhaps more significantly, the automation of the laboratory allows and encourages increasingly complex and sophisticated analysis. Differences between sample results once tested using manual t tests can be further analyzed with multiple regressions and nonparametric statistical analyses. More subtle differences can be detected, leading to higher levels of QC, previously neglected drug discoveries, more tightly controlled human and animal testing, and generally higher levels of sophistication. The sophistication of these analyses and the tight controls and analyses of increasingly minute differences, however, increase the importance of the accuracy and reliability of laboratory data, and the tool for assuring that accuracy and reliability is GLP.

As we rely more on laboratory results and as we more carefully examine minor differences in those results, the GLPs will play an increasingly important role.

SUMMARY

In examining the history of the GLPs and the application of the regulation to a diversity of environments, including non-GLP environmental labs, clinical and R&D settings, and the QC role of the GLPs, one set of factors clearly emerges. The GLPs represent not only state-of-the-art thinking in laboratory management and control, but serve as a living collection of documents that can adapt and evolve to changes in laboratory operations, automation, and application. In the foreseeable future this flexibility and adaptability are likely to continue to be the most significant features of the GLPs.

Six general predictions have been derived from expectations in the changing world of automation, laboratory function, and regulation.

That automation will continue to grow in both extent and degree, evolving as the determining factor in lab operations.

That this automation will lead to the development of additional lab-in-a-box, self-contained robotic analytical devices

That a countertrend will lead to the development of large, multiuser centralized laboratories capable of performing a variety of complex tests on a number of separate data sets in a fully compliant environment

That both of these trends—toward increased automation and increasingly complex labs—will result in a growing regulatory emphasis on data quality, including assurances of the integrity of the data input process

That OSHA is likely to increase its involvement in worker safety in laboratory environments unless a GLP revision places this responsibility on the shoulders of FDA rather than OSHA investigators

That as a result of these changes and of the increasing reliance on laboratory quality roles and the increasing sophistication of testing that automation permits, the GLPs will continue to play a key role in the pharmaceutical industry and processes

These six predictions combine to make clear that the GLPs will continue to be one of the most important set of guidelines ever developed by the FDA. The GLPs have proven to be a useful template for a variety of kinds of laboratories, performing a variety of missions, in a wide range of environments, over a period of significant evolution of function.

A fundamental reason for such importance is the ability of the GLPs to evolve in that changing environment. Because the GLPs provide principle rather than detail, because they focus on guidance rather than requirement, and because they reflect approach over specific solutions they remain relevant despite incredible changes in the nature and role of the modern laboratory.

Predicting the future of the GLPs and of the changes affecting laboratories and their regulation isn't difficult; no tea leaves or crystal balls are required. By examining the current relatives—the entrails of selected sheep from the herd—we can quickly define the major trends: more automation; increased regulation, particularly in the area of safety; increased centralization; and a continued reliance on the principles and guidelines of the GLPs. The predictions are safe because the trends are firmly rooted in the present and the past, and the past, present, and foreseeable future of laboratory control is defined by the GLPs.

Bibliography

Allinson C. Information systems audit trails in legal proceedings as evidence. Computers Sec 20.5: 409–421, 2001.

Alter S. *Decision Support Systems: Current Practice and Continuing Challenges*. Reading, MA: Addison-Wesley, 1980.

American Institute of Certified Public Accountants. Audit Approaches for a Computerized Inventory System. New York: AICPA, 1980.

American Institute of Certified Public Accountants. Audit Considerations in Electronic Fund Transfer Systems. New York: AICPA, 1979.

American Institute of Certified Public Accountants. Guidelines to Assess Computerized General Ledger and Financial Reporting Systems for Use in CPA Firms. New York: AICPA, 1979.

Andersen A. A Guide for Studying and Evaluating Internal Controls. Chicago: Arthur Andersen, 1978.

Anderson CA. Approach to data processing auditing. In: *The Interpreter*. Durham, NC: Insurance Accounting and Statistical Association, April 1975, pp. 23–26.

Arthus LJ. *Measuring Programmer Productivity and Software Quality*. New York: Wiley, 1985.

Aron J. The Program Development Process, Part II: The Programming Team. Reading, MA: Addison-Wesley, 1983.

Automated Laboratory Standards: Current Automated Laboratory Data Management Practices, Final. June 1990.

Automated Laboratory Standards: Evaluation of the Standards and Procedures Used in Automated Clinical Laboratories, Draft. May 1990.

Automated Laboratory Standards: Evaluation of the Use of Automated Financial System Procedures, Final. June 1990.

Automated Laboratory Standards: Good Laboratory Practices for EPA Programs, Draft. June 1990.

Automated Laboratory Standards: Survey of Current Automated Technology, Final. June 1990.

Bar-Hava N. Training EDP auditors. *Info Mgt* (4):30–42, 1981.

Basden A, Clar EM. Data integrity in a general practice computer system (CLINICS). *Internat J Biomed Comput* 11:511–519, 1980.

Bennett JL, ed. *Building Decision Support Systems*. Reading, MA: Addison-Wesley, 1983.

Bigger staff, TJ, Perlins AJ, eds. *Software Reusability*. vol. 2, Applications and Experience. New York: ACM Press/Addison-Wesley, 1989.

Board of Governors of the Federal Reserve System. Electronic Fund Transfers. *Regulation E* (12 CFR Part 205). effective March 30, 1979 (as amended effective May 10, 1980.)

Boehm BW. *Software Engineering Economics*. Englewood Cliffs, NJ: Prentice-Hall, 1982.

Bonczek RH, Holsapple CW, Whinston AB. Foundations of Decision Support Systems. New York: Academic, 1981.

Brooks F. *The Mythical Man-Month*. Reading, MA: Addison-Wesley, 1975.

Bronstein RJ. The concept of a validation plan. *Drug Info J* 20: 37–42, 1986.

Brown EH. Procedures and their documentation for a LIMS in a regulated environment. In: RD McDowall, ed. *Laboratory Information Management Systems*. Wilmslow, UK: Sigma, 1987, pp. 346–358.

BURMARC. Pocket sized meters. Soap, Perf, Cosmet 71.5: 59, 1998.

Callahan JJ. *Needed: Professional Management in Data Processing*. Englewood Cliffs, NJ: Prentice-Hall, 1983.

Caputo CA. Managing the EDP audit function. *COM-SAC* 8(1): A-1–A-8, Jan. 1981.

Card DN, Glass RL. *Measuring Software Design Quality*. Englewood Cliffs, NJ: Prentice-Hall, 1990.

Casti JL. Paradigms Lost. New York: William Morrow, 1989.

Charlette, RN. *Software Engineering Risk Analysis and Management*. New York: McGraw-Hill, 1989.

Cheney PH. Educating the computer audit specialist. *EDP Aud* 9–15, fall 1980.

Clinical Laboratory Improvement Amendments of 1988. P.L. 100-578. Oct. 31, 1988.

College of American Pathologists. *Standards for Laboratory Accreditation*. Skokie, IL: Commission on Laboratory Accreditation, 1988.

Connor JE, DeVos BH. Guide to Accounting Controls—Establishing, Evaluating and Monitoring Control Systems. Boston: Warren, Gorham & Lamont, 1980.

Cooke JE, Drury, DH. Management Planning and Control of Information Systems. Hamilton, Ontario: Society of Management Accountants of Canada, April 1980.

CRS releases enhanced POLARA™ software for the drug development. Can Corp News Jan. 29, 2001.

Data Acquisition Telecommunications Local Area Networks. *1983 Data Book*. Sunnyvale, CA: Advanced Micro Devices, 1983.

DataMirror LiveAudit™ for FDA-regulated organizations supports leading ERP solutions. Can Corp News June 5, 2001.

DataMirror unveils data integration solution for FDA-regulated organizations. Can Corp News May 29, 2001.

Datapro Research. *Datapro Reports on Information Security*. Delran, NJ: McGraw-Hill, 1989.

Davis SM. *Future Perfect*. Reading, MA: Addison-Wesley, 1987.

Department of Transportation. Procedures for transportation workplace drug testing programs: Final rule. *Fed Reg* 54 (230): 49854–49884, Dec. 1, 1989.

Dessy RE. The Electronic Laboratory. Washington, DC: American Chemical Society, 1985.

DeMarco T. *Controlling Software Projects*. New York: Yourdon, 1982.

DeMarco T, Lister T. *Software State-of-the-Art: Selected Papers*. New York: Dorset, 1991.

Dice B. operations manager, telephone interview, Hyattsville, MD, April 25, 1990.

U.S. Department of Defense. Defense System Software Development. DOD-STD-2167A. Feb. 29, 1988.

Drug Information Association. Computerized Data Systems for Nonclinical Safety Assessment: Current Concepts and Quality Assurance. Maple Glen, PA: Drug Information Allocation, 1980.

Dunn R. *Software Defect Removal*. New York: McGraw-Hill, 1984.

Dunn R, Ullman R. *Quality Assurance for Computer Software*. New York: McGraw-Hill, 1982.

EDP Auditors Foundation. Control Objectives—1980. Streamwood, IL: EDP Auditors Foundation, 1980.

Figarole PL. Computer software validation techniques. DIA Conference on Computer Validation, Jan. 21–23, 1985.

First state drug safety evaluation lab established. Xinhua News Ag Aug. 28, 2000.

Forsyth A. An approach to audit an on-line system. *COM-SAC* 7 (1) A-1–A-4, Jan. 1980.

Freedman D, Weinberg G. *Handbook of Walkthroughs, Inspections and Technical Reviews*. Boston: Little, Brown, 1982.

Gallegos F, Bieber D. What every auditor should know about computer information systems. Available as accession number 130454 from the General Accounting Office (GOA) and reprinted from p. 1–11 in EDP. *Auditing*. New York: Auerbach, 1986.

Gardner E. Computer dilemma: Clinical access vs. confidentiality. *Mod Healthcare*. 32–42, Nov. 3, 1989.

Gardner E. Secure passwords and audit trails. sidebar. *Mod Healthcare* 33, Nov. 3, 1989.

Gardner E. System assigns passwords, beeps at security breaches. sidebar. *Mod Healthcare* 34, Nov. 3, 1989.

Gardner E. System opens access to physicians, restricts it to others. sidebar. *Mod Healthcare* 38, Nov. 3, 1989.

Gardner E. "Borrowed" passwords borrow trouble. sidebar. *Mod Healthcare* 42, Nov. 3, 1989.

Gardner E. Recording results of AIDS tests can be a balancing act. sidebar. *Mod Healthcare* 40, Nov. 3, 1989.

Garwood RM. FDA's viewpoint on inspection of computer systems.

Course notes from Computers in Process Control: FDA Course, March 19–23, 1984.
Geffkin C. Does your lab measure up? Global Cos Ind 169.3: 13–31, 2001.
Gild T. *Principles of Software Engineering Management*. Reading, MA: Addison-Wesley, 1988.
Gilhooley IA. Improving the relationship between internal and external auditors. *J Acc EDP* 1 (1): 4–9, Spring 1985.
Glover DE, Hall RG, Coston AW, Trilling RJ. Validation of data obtained during exposure of human volunteers to air pollutants. *Computers Biomed Res* 15(3):240–249, 1982.
Good Automated Laboratory Practices (GALPs). Dec. 28, 1990.
Goren LJ. Computer system validation, part II. *BioPharm* 38–42, Feb. 1989.
Guynes S. EFTS impact on computer security. *Computers* Sec 2 (1): 73–77, 1983.
Hagland M. Confidence and confidentiality. *Health Mgt Techn* 18.12: 20–24, 1997.
Halper S, Davis D, Glen C, O'Neil-Dunne JP, Pfau PR. In: *Handbook of EDP Auditing*. vol. 28, Testing techniques for computer-based system. 1985, 1–26.
Hanna meters are portable. *Mfg Chem* 66.11: 53, 1995.
Hatley D, Imtiaz P. *Strategies for Real-Time System Specification*. New York: Doreset, 1987.
Haworth R, Pilling AM. The PCR assay in the preclinical safety evaluation of nucleic acid medicines. Human Experi Toxicol 19.5 (200): 267–276.
Hayes JR. The Complete Problem Solver. Philadelphia: Franklin Institute, 1981.
Hayes-Roth F, Waterman DA, Lenat DB. *Building Expert Systems*. Reading, MA: Addison-Wesley, 1983.
Henry C. Food and Drug Administration. Anal Chem 71.5: 127–130, 1999.
Hetzel W. *The Complete Guide to Software Testing*. 2nd ed. Wellesley, MA: QED Information Sciences, 1988.
Hitachi adds to applications. *Mfg Chem* 66.11: 53, 1995.
Hoaglin DC, Mosteller F, Turkey JW. *Understanding Robust and Exploratory Data Analysis*. New York: Wiley, 1983.
Hubbert J. Data base concepts. *EDP Aud* spring 1980.

Humphrey WS. *Managing the Software Process*. Reading, MA: Addison-Wesley, 1989.

Ion measurement made easier through automation. J Environ Health 58.8: 30–31, 1996.

Jackson M. *System Development*. Englewood Cliffs, NJ: Prentice Hall, 1983.

Jacobs JD, Walker MJ. Introducing the environmental data police in the decade of data and data quality. J Environ Law Prac 4.1: 47–53, 1996.

Johnson CD. *Process Control Instrumentation Technology*. 2nd ed. New York: Wiley, 1982.

Kull D. Demystifying ergonomics. *Computer Dec* Sept. 1984.

Kuong JF. Approaches to justifying EDP controls and auditability provisions. *COM-SAC* 7 (2): A-1–A-8, July 1980.

Kuong JF. Audit and Control of Advanced/On-Line Systems. MAP-7. Management Advisory Publications, 1980.

Kuong JF. Auditor involvement in system development and the need to develop effective, efficient, secure, auditable and controllable systems. keynote speech at the First Regional EDP Auditors Conference, Tel-Aviv, Israel, June 3, 1982.

Kuong JF. Controls for Advanced/On-line/Data-Base Systems. vols. 1 and 2. Management Advisory Publications, 1985.

Kuong JF. Managing the EDP audit function. paper presented at the First Regional EDP Auditors Conference, Tel-Aviv, Israel, May 29–June 3, 1982.

Kuong JF. Organizing, managing and controlling the EDP auditing function. seminar text, Management Advisory Publications, 1980.

Luarel B, ed. *The Art of Human–Computer Interface Design*. Reading, MA: Addison-Wesley, 1990.

Litecki CR, McEnroe JE. EDP audit job definitions: How does your staff compare? *Internal Aud* 57–61, April 1981.

Macchiaverna PR. Auditing Corporate Data Processing Activities. New York: Conference Board, 1980.

MALVERN. Sizing up to GMP. Soap, Perf Cosmet 69.9: 45, 1996.

Marks RC. Performance appraisal of EDP auditors. speech given at the 11th Conference on Computer Auditing, Security and Control, ATC/IIA, New York, May 4–8, 1981.

Marks WE. Evaluating the information systems staff. *Info Syst News* Dec. 24, 1984.

Mason RO, Swanson EB. *Measurement for Management Decision*. Reading, MA: Addison-Wesley, 1981.

Mattes DC. LIMS and good laboratory practice. In RD McDowall, ed. *Laboratory Information Management Systems*. Wilmslow, UK: Sigma, 1987, pp. 332–345.

McClure C. *CASE Is Software Automation*. Englewood Cliffs, NJ: Prentice-Hall, 1989.

McClure C. *Managing Software Development and Maintenance*. New York: Van Nostrand Reinhold, 1981.

McDowall RD, ed. *Laboratory Information Management Systems*. Wilmslow, UK: Sigma, 1987.

McMenamin S, Palmer J. *Essential Systems Analysis*. Englewood Cliffs, NJ: Yourdin Press/Prentice-Hall, 1984.

Megargle R. Laboratory information management systems. *Anal Chem* 61(9): 612A–621A, 1989.

Merrer RJ, Berthrong PG. Academic LIMS: Concept and practice. *Amer Lab* 21(3): 36–45, 1989.

Mills H, Linger R, Hevner A. *Principles of Information Systems Analysis and Design*. New York: Academic, 1986.

Mills ME. Computer-based health care data and the Health Insurance Portability and Accountability Act: Implications and informatics. Policy, Politics, and Nurs Prac 2.1: 33–38, 2001.

Morris R III. The internal auditors and data processing. *Internal Aud* Aug. 1978.

Musa J, Iannino A, Okumoto K. *Software Reliability: Measurement, Prediction, Application*. New York: McGraw-Hill, 1987.

Meyers W. Build defect-free software, Fagan urges. *Computer IEEE* Aug. 1990.

National Computer Security Center. Glossary of Computer Security. U.S. Dept. of Defense, NCSC-TG-004-88, version 1, 1988.

Normile D. Court backs lab's safety practices. *Science* 292.5514: 30, 2001.

Nortran arrhythmia drug demonstrates oral bioavailability. Can Corp News June 21, 2001.

Office of Information Resources Management. EPA LIMS Functional Specification. Washington, DC: Environmental Protection Agency, March 1988.

Office of Information Resources Management. *EPA System Design and Development Guidance*. vols. A, B, and C. Washington, DC: U.S. Environmental Protection Agency, 1989.

Office of Information Resources Management. *Survey of Laboratory Automated Data Management Practices*. Research Triangle Park, NC: U.S. Environmental Protection Agency, 1989.

Office of Management and Budget. *Guidance for Preparation and Submission of Security Plans for Federal Computer Systems Containing Sensitive Information*. OMB bulletin no. 88-16. Washington, DC: Office of Management and Budget, July 6, 1988.

Parikh G. The politics of software maintenance. *Infosystems* Aug. 1984.

Perry WE, Kuong JF. EDP Risk Analysis and Controls Justification. Management Advisory Publications, 1981.

Perry WE, Kuong JF. Generalized Computer Audit Software—Selection and Application. MAP-14. Management Advisory Publications, 1980.

Phipps G. Practical application of software testing. Computer Sciences Corporation, presented Oct. 8, 1986.

Pinkus KV. Financial auditing and fraud detection: Implications for scientific data audit. *Acct Res* 1:53–70. 1989.

Pressman RS. *Making Software Engineering Happen*. Englewood Cliffs, NJ: Prentice-Hall, 1988.

Products-information management. *Lab Prac*, 38(5):87–91, May 1989.

Reilly RF, Lee JA. Developing an EDP audit function: A case study. *EDPACS Newsl* 1–10, May 1981.

Reinmann BC, Warren AD. User-oriented criteria for the selection of Dds software. Commun ACM 28 (2): 166–179, Feb. 1985.

Report evaluates lab-safety trends. R and D 37.12: 13, 1995.

Romano CA. Privacy, confidentiality, and security of computerized systems: The nursing responsibility. *Computers Nurs* 99–104, May/June 1987.

Sandowdki C, Lawler G. A relational data base management system for LIMS. *Amer Lab* 21(3):70–79, March 1989.

SARTORIOUS. Self-calibrating balances for quality assurance systems. Pharm Mfg Rev 8.5: 38, 1996.

SARTORIOUS. Balances allow self-calibration. Mfg Chem 67.10: 87, 1996.

Schindler M. *Computer-Aided Software Design*. New York: Wiley, 1990.

Schroeder FJ. Developments in consumer electronic fund transfers. *Fed Res Bull* 69(6):395–403, 1983.

Schulmeyer GG. *Zero Defect Software*. New York: McGraw-Hill, 1990.

SignalGene signs an agreement with Allied Clinical Research for pharmacogenomics services. Can Corp News Oct. 30, 2000.

Smith MR. *Commonsense Computer Security*. London: McGraw-Hill, 1989.

Ince D, ed. *Software Quality and Reliability*. London: Chapman and Hall, 1991.

Tussing RT, Hellms GL. Training computer audit specialists. *J Acct* 71–74, July 1980.

U.S. Dept. Defense. Defense System Software Development Data Item Description for the Software Design Document. DI-MCCR-80012A. Washington, DC: GPO, Feb. 29, 1988.

U.S. Dept. of Health and Human Services. Mandatory guidelines for federal workplace drug testing programs: Final guidelines. *Fed Reg* 53 (69): 11969–9610, April 11, 1988.

U.S. drug approval process slows. Australian Feb. 7, 2001.

U.S. General Accounting Office. Evaluating the Acquisition and Operation of Information Systems. Washington, DC: General Accounting Office, 1986.

U.S. General Accounting Office. Bibliography of FAO Documents, *ADP, IRM< & Telecommunications 1986*, Washington, DC: General Accounting Office, 1987.

Vasarhelyi MA, Pabst CA, Daley I. Organizational and career aspects of the EDP audit function. *EDP Aud* 35–43, 1980,

Vogel G. FDA report scores chimp research lab. Science 286.5443: 1269–1270, 1999.

Weinberg SB. *System Validation Checklist*. 1988.

Weinberg SB. *Testing Protocols for the Blood Processing Industry*. 1989.

Weinberg SB. *System Validation Standards*. Dubuque, IA: Kendall/Hunt, 1990.

What Every Auditor Should Know About DP. New York: Data Processing Management Service, Auerbach, Portfolio 3-09-03, 1975, p. 14.

What's new. Ceram Ind 151.9: 74, 2001.

Yourdon E. *Structured Walkthroughs*. New York: Yourdon, 1982.

Index